PAR

FOUNDATIONS OF SOFT CASE-BASED REASONING

WILEY SERIES ON INTELLIGENT SYSTEMS

Editors: James S. Albus, Alexander M. Meystel, and Lotfi A. Zadeh

FOUNDATIONS OF SOFT CASE-BASED REASONING

SANKAR K. PAL
Indian Statistical Institute

SIMON C. K. SHIU
Hong Kong Polytechnic University

Ⓦ **WILEY-INTERSCIENCE**

A JOHN WILEY & SONS, INC., PUBLICATION

Library of Congress Cataloging-in-Publication Data:

Pal, Sankar
   Foundations of soft case-based reasoning / Sankar Pal, Simon Shiu
      p.   cm. – (Wiley series on intelligent systems)
"A Wiley-Interscience publication."
Includes bibliographical references and index.
ISBN 0-471-08635-5
   1. Soft computing. 2. Case-based reasoning   I. Shiu, Simon C. K.
II. Title.   III. Series.
QA76.9.S63 S55 2004
006.3–dc22   2003021342

Printed in the United States of America

10 9 8 7 6 5 4 3 2 1

To
Anshu, Arghya, and Amita
SKP

Pak Wah, Yee Shin, and Mei Yee
SCKS

CONTENTS

5 CASE-BASE MAINTENANCE   161

6 APPLICATIONS   201

FOREWORD

To say that *Foundations of Soft Case-Based Reasoning* (FSCBR for short) is a work of great importance is an understatement. Authored by prominent information scientists, Professors S. K. Pal and S. Shiu, it breaks new ground in case-based reasoning and is likely to be viewed in retrospect as a milestone in its field.

Case-based reasoning (CBR) is a body of concepts and techniques that touch upon some of the most basic issues relating to knowledge representation, reasoning, and learning from experience. The brainchild of Janet Kolodner and others, it was born in the early 1980s. I witnessed its birth and followed with great interest its evolution and coming of age. But when I tried to develop a better understanding of CBR, I encountered a problem. The core methods of CBR were not powerful enough to address the complex concepts and issues that had to be dealt with. A case in point is the concept of similarity, a concept that plays a pivotal role in CBR. The late Amos Tversky—a brilliant cognitive scientist—had defined the degree of similarity of objects A and B as a ratio whose numerator is the number of features that A and B have in common, and whose denominator is the total number of features. The problem with this definition is that it presupposes (1) that features are bivalent, whereas in most realistic settings at least some of the features are multivalent; and (2) that features are of equal importance, which in most instances is not the case.

It is beyond question that impressive progress has been made since the early days of CBR in our understanding of concepts such as similarity, relevance, and materiality. But a basic question that motivated the work of Professors Pal and Shiu is: Is it possible to achieve a quantum jump in capabilities of CBR systems through the use of traditional methods of computing and reasoning? In effect, FSCBR may be viewed as a negative answer to this question. But more important, FSCBR is a constructive answer because the authors demonstrate that a quantum jump is achievable through the development of what they refer to as soft case-based reasoning (SCBR).

SCBR is based on soft computing, a computing methodology that is a coalition of methodologies which collectively provide a foundation for the conception, design, and utilization of intelligent systems. The principal members of the coalition are fuzzy logic, neurocomputing, evolutionary computing, probabilistic computing, chaotic computing, rough set theory, self-organizing maps, and machine learning. An essential tenet of soft computing is that, in general, superior results can be achieved by employing the constituent methodologies of soft computing in combination rather than in a stand-alone mode. A combination that has achieved high visibility is known as neuro-fuzzy. Another combination is neuro-fuzzy-genetic.

To see SCBR in a proper perspective, a bit of history is in order. In science, as in most other realms of human activity; there is a tendency to be nationalistic—to commit oneself to a particular methodology, M, and march under its banner, in the belief that it is M and only M that matters. The well-known Hammer Principle: When the only tool you have is a hammer, everything looks like a nail; and my Vodka Principle: No matter what your problem is, vodka will solve it, are succinct expressions of the one-size-fits-all mentality that underlies nationalism in science. Although it is obvious that a one-size-fits-all mentality in science is counter-productive, the question is: What can be done to counter it?

A step in this direction was taken at UC Berkeley in 1991, with the launching of what was called the Berkeley Initiative in Soft Computing (BISC). The principal tenet of BISC is that to come up with effective tools for dealing with the complex problems that arise in the conception, design, and utilization of intelligent systems, it is imperative to marshal all the resources at our disposal by forming a coalition of relevant methodologies and developing synergistic links between them. An important concomitant of the concept of soft computing is that students should be trained to feel at home with all or most of the constituent methodologies of soft computing and to be able to use them both singly and in combination, depending on the nature of the problem at hand.

This is the spirit that underlies the work of Professors Pal and Shiu. They start with an exposition of traditional methods but then cross into new territory and proceed to develop soft case-based reasoning as a unified theory that exploits the wide diversity of concepts and techniques drawn from constituent methodologies of soft computing. To make their work self-contained, the authors include in FSCBR succinct and insightful expositions of the basics of fuzzy logic, neurocomputing, genetic algorithms, rough set theory, and self-organizing maps.

What Professors Pal and Shiu have done is a truly remarkable accomplishment. The authors had to master a wide spectrum of concepts and techniques within soft computing and apply their expertise to the development of a comprehensive theory of soft case-based reasoning, a theory that is certain to have a wide-ranging impact in fields extending from diagnostics and data mining to law, medicine, and decision analysis.

FSCBR is ideally suited both as a reference and as a text for a graduate course on case-based reasoning. Beyond CBR, FSCBR is must reading for anyone who is interested in the conception, design, and utilization of intelligent systems. The authors, Professors Pal and Shiu, and the publisher, John Wiley, deserve loud applause for producing a book that is a major contribution not only to case-based reasoning but, more generally, to the conception, design, and utilization of intelligent systems.

LOTFI ZADEH
Berkely, CA
September 22, 2003

PREFACE

There has been a spurt of activity during the last decade to integrate various computing paradigms, such as fuzzy set theory, neural networks, genetic algorithms, and rough set theory, for generating more efficient hybrid systems that can be classified as soft computing methodologies. Here the individual tool acts synergistically, not competitively, in enhancing the application domain of the other. The purpose is to develop flexible information-processing systems that can exploit the tolerance for imprecision, uncertainty, approximate reasoning, and partial truth in order to achieve tractability, robustness, low solution cost, and close resemblance to human decision making. The soft computing paradigm provides a foundation for the conception and design of high-MIQ (machine IQ) systems and forms a basis for future-generation computing technology. The computational theory of perceptions (CTP) described by Zadeh, with perceptions being characterized by fuzzy granularity, plays a key role in performing tasks in a soft computing framework. Tremendous efforts are being made along this line to develop theories and algorithms on the one hand, and to demonstrate various applications on the other, considering its constituting tools both individually and in different combinations.

Case-based reasoning (CBR) is one such application area where soft computing methodologies have had a significant impact during the past decade. CBR may be defined as a model of reasoning that incorporates problem solving, understanding, and learning, and integrates all of them with memory processes. These tasks are performed using typical situations, called *cases*, already experienced by a system. A case may be defined as a contextualized piece of knowledge representing an experience that teaches a lesson fundamental to achieving the goals of the system. The system learns as a by-product of its reasoning activity. It becomes more efficient and more competent as a result of storing the past experience of the system and referring to earlier cases in later reasoning. Unlike a traditional knowledge-based system, a case-based system operates through a process of remembering one or a small set of concrete instances or cases and basing decisions on comparisons between the new and old situations. Systems based on this principle are finding widespread applications in such problems as medical diagnosis and legal interpretation where the knowledge available is incomplete and/or evidence is sparse.

Four prime components of a CBR system are retrieve, reuse, revise, and retain. These involve such basic tasks as clustering and classification of cases, case selection and generation, case indexing and learning, measuring case similarity, case retrieval and inference, reasoning, and rule adaptation and mining. The use of soft computing tools in general, and fuzzy logic and artificial neural networks in performing these tasks in particular, has been well established for more than a decade. The primary roles of these tools are in handling ambiguous, vague, or ill-defined information or concepts, learning and adaptation of intractable cases or classes, searching for optimal parameters, and computing with granules (clumps of similar objects or cases) for speedy computation. CBR systems that integrate these characteristics in various combinations for developing efficient methodologies, algorithms, and knowledge-based networks for various real-life decision-making applications have also been developed.

This book provides a treatise in a unified framework describing how soft computing techniques can be used in building and maintaining case-based systems. The book is structured according to the four major phases of the problem-solving life cycle of a CBR system—representation and indexing of cases, case selection and retrieval, case adaptation, and case-base maintenance—and provides a solid foundation with a balanced mixture of theory, algorithm, and application. Examples are provided wherever necessary to make the concepts more clear. Various real-life applications are presented in a comprehensive manner for the benefit of practitioners.

For the convenience of readers, the basic theories, principles, and definitions of fuzzy sets, artificial neural networks, genetic algorithms, and rough sets are provided in the appendixes. A comprehensive bibliography is provided for each chapter. A sizable portion of the text has been unified from previously published work of the authors.

The book, which is unique in character, will be useful to graduate students and researchers in computer science, electrical engineering, system science, and information technology as both a textbook and a reference book for some parts of the curriculum. Researchers and practitioners in industry and R&D laboratories working in such fields as system design, control, pattern recognition, data mining, vision, and machine intelligence will also be benefited.

The text is organized in six chapters. In Chapter 1 we provide an introduction to CBR system together with its various components and characteristic features and an example of building a CBR system. This is followed by a brief description of the soft computing paradigm, an introduction to soft case-based reasoning, and a list of typical CBR tasks for soft computing applications.

Chapter 2 highlights problems of case representation and indexing. Here we describe, first, traditional methods of case representation: relational, object-oriented, and predicate representation. This is followed by a method of case knowledge representation using fuzzy sets, examples of determining reducts from a decision table using rough sets, and a methodology of prototypical case generation in a rough-fuzzy framework. The significance of granular computing is demonstrated. Some experimental results on case generation are also provided for large data sets. Finally, some case indexing methods using a traditional approach, a Bayesian model, and neural networks are described.

Chapter 3 deals with the tasks of case selection and retrieval. We begin with problems in constructing similarity measures by defining a few well-known similarity measures in terms of distance, followed by the relevance of the concept of fuzzy similarity between cases and some methods of computation. Methods of computing feature weights using classical, neural, and genetic algorithm–based approaches are then discussed. Finally, various methodologies of case selection and retrieval in neural, neuro-fuzzy, and rough-neural frameworks are described. Here both layered network and self-organizing maps are considered for learning in supervised and unsupervised modes, and experimental results demonstrating the features are given.

Issues of case adaptation are handled in Chapter 4. After explaining some conventional strategies—reinstantiation, substitution and transformation—and a few methods based on them, various ways of using fuzzy decision trees, multilayer perceptrons, Bayesian models, and support vector machines for case adaptation are presented. We explain how discrepancy vectors can be used as training examples for determining the amount of adjustment needed to modify a solution. The use of genetic algorithms in this regard is also discussed.

Chapter 5 is concerned with problems of case-base maintenance. We first explain different characteristic properties that need to be assured through qualitative and quantitative maintenance. Then two methods of case-base maintenance using fuzzy-rough and fuzzy integral approaches are described. Tasks such as mining adaptation rules, adjustment through reasoning, selecting cases and updating the case base; and such concepts as case coverage and reachability, fuzzy integrals, and case-base competence are explained in detail through example computations. Some experimental results are also provided, as in earlier chapters.

Finally, some real-life applications of soft case-based reasoning systems are presented in a comprehensive manner in Chapter 6, together with their significance and merits. These include Web access path prediction, oceanographic forecasting, medical diagnosis, legal inference, property valuation, bond rating, color matching, and fashion shoe design.

We take this opportunity to thank John Wiley & Sons for its initiative and encouragement. We owe a vote of thanks to Ms. Yan Li and Mr. Ben Niu of the Department of Computing, Hong Kong Polytechnic University, for their tireless endeavors in providing remarkable assistance while preparing the manuscript, as well as to colleagues at the Machine Intelligence Unit, Indian Statistical Institute, Calcutta, and Professor Tharam S. Dillon, La Trobe University, Melbourne, for their cooperation at various stages. Financial support from Hong Kong Polytechnic University, grants HZJ90, GT377, and APD55, and RGC grant BQ496, is also gratefully acknowledged. The project was initiated when Professor Pal was a Visiting Professor at the Hong Kong Polytechnic University, Hong Kong, during 2000–2001. The names of the authors are arranged alphabetically, signifying their equal contribution.

SANKAR K. PAL
SIMON C. K. SHIU
June 29, 2003

ABOUT THE AUTHORS

Sankar K. Pal is a *Professor* and *Distinguished Scientist* at the Indian Statistical Institute, Calcutta, and the *Founding Head* of the Machine Intelligence Unit. He received the M. Tech. and Ph.D. degrees in radio physics and electronics in 1974 and 1979, respectively, from the University of Calcutta. In 1982 he received another Ph.D., in electrical engineering, along with a DIC from Imperial College, University of London. During 1986–1987 he worked at the University of California, Berkeley and the University of Maryland, College Park, as a *Fulbright Post-doctoral Visiting Fellow*; and during 1990–1992 and in 1994 at the NASA Johnson Space Center, Houston, Texas as an *NRC Guest Investigator*. He has been a *Distinguished Visitor* of the IEEE Computer Society (United States) for the Asia-Pacific Region since 1997, and held several visiting positions in Hong Kong and Australian universities during 1999–2003.

Professor Pal is a *Fellow* of the IEEE and the International Association for Pattern Recognition in the United States, and all the Third World Academy of Sciences in Italy, four National Academies for Science/Engineering in India. His research interests include pattern recognition and machine learning, image processing, data mining, soft computing, neural nets, genetic algorithms, fuzzy sets and rough sets, Web intelligence, and bioinformatics. He is a coauthor of 10 books and about 300 research publications. He has received the *1990 S. S. Bhatnagar Prize* (the most coveted award for a scientist in India) and many other prestigious awards in India and abroad, including the *1993 Jawaharlal Nehru Fellowship, 1993 Vikram Sarabhai Research Award, 1993 NASA Tech Brief Award, 1994 IEEE Transactions on Neural Networks Outstanding Paper Award (USA), 1995 NASA Patent Application Award, 1997 IETE–Ram Lal Wadhwa Gold Medal, 1998 Om Bhasin Foundation Award, 1999 G. D. Birla Award for Scientific Research, 2000 Khwarizmi International Award (first winner)* from the Islamic Republic of Iran, the *2001 INSA–Syed Husain Zaheer Medal,* and the *FICCI Award 2000–2001 in Engineering and Technology.*

Professor Pal has been or is serving as *Editor, Associate Editor,* and *Guest Editor* of *IEEE Transactions on Pattern Analysis and Machine Intelligence, IEEE Transactions on Neural Networks, IEEE Computer, Pattern Recognition Letters, Neurocomputing, Applied Intelligence, Information Sciences, Fuzzy Sets and Systems, Fundamenta Informaticae,* and *International Journal of Computational Intelligence and Applications*; and is a *Member of the Executive Advisory Editorial Board, IEEE Transactions on Fuzzy Systems, International Journal on Image and Graphics,* and *International Journal of Approximate Reasoning.*

Simon C. K. Shiu is an Assistant Professor in the Department of Computing, Hong Kong Polytechnic University, Hong Kong. He received an M.Sc. degree in computing science from the University of Newcastle Upon Tyne, United Kingdom, in 1985, a M.Sc. degree in business systems analysis and design from City University, London in 1986, and a Ph.D. degree in computing in 1997 from Hong Kong Polytechnic University. Between 1985 and 1990 he worked as a system analyst and project manager in several business organizations in Hong Kong. His current research interests include case-based reasoning, machine learning, and soft computing. He has *co-guest edited* a special issue on soft case-based reasoning for the journal *Applied Intelligence.* Dr. Shiu is a member of the British Computer Society and the IEEE.

CHAPTER 1

INTRODUCTION

1.1 BACKGROUND

The field of case-based reasoning (CBR), which has a relatively young history, arose out of the research in cognitive science. The earliest contributions in this area were from Roger Schank and his colleagues at Yale University [1,2]. During the period 1977–1993, CBR research was highly regarded as a plausible high-level model for cognitive processing. It was focused on problems such as how people learn a new skill and how humans generate hypotheses about new situations based on their past experiences. The objectives of these cognitive-based researches were to construct decision support systems to help people learn. Many prototype CBR systems were built during this period: for example, Cyrus [3,4], Mediator [5], Persuader [6], Chef [7], Julia [8], Casey, and Protos [9]. Three CBR workshops were organized in 1988, 1989, and 1991 by the U.S. Defense Advanced Research Projects Agency (DARPA). These formally marked the birth of the discipline of case-based reasoning. In 1993, the first European workshop on case-based reasoning (EWCBR-93) was held in Kaiserslautern, Germany. It was a great success, and it attracted more than 120 delegates and over 80 papers. Since then, many international workshops and conferences on CBR have been held in different parts of the world, such as the following:

- Second European Workshop on CBR (EWCBR-94), Chantilly, France
- First International Conference on CBR (ICCBR-95), Sesimbra, Portugal

Foundations of Soft Case-Based Reasoning.   By Sankar K. Pal and Simon C. K. Shiu
ISBN 0-471-08635-5   Copyright © 2004 John Wiley & Sons, Inc.

- Third European Workshop on CBR (EWCBR-96), Lausanne, Switzerland
- Second International Conference on CBR (ICCBR-97), Providence, Rhode Island
- Fourth European Workshop on CBR (EWCBR-98), Dublin, Ireland
- Third International Conference on CBR (ICCBR-99), Seeon Monastery, Munich, Germany
- Fifth European Workshop on CBR (EWCBR-00), Trento, Italy
- Fourth International Conference on CBR (ICCBR-01), Vancouver, Canada
- Sixth European Conference on CBR (ECCBR-02), Aberdeen, Scotland
- Fifth International Conference on CBR (ICCBR-03), Trondheim, Norway
- Seventh European Conference on CBR (ECCBR-04), Madrid, Spain

Other major artificial intelligence conferences, such as ECAI (European Conference on Artificial Intelligence), IJCAI (International Joint Conference on Artificial Intelligence), and one organized by the AAAI (American Association for Artificial Intelligence), have also had CBR workshops as part of their regular programs. Recently, CBR has drawn the attention of researchers from Asia, such as the authors of this book, from countries such as Hong Kong and India.

The rest of this chapter is organized as follows. In Section 1.2 we describe the various components and features of CBR. The guidelines and advantages of using CBR are explained in Sections 1.3 and 1.4, respectively. In Section 1.5 we address the tasks of case representation and indexing, and in Section 1.6 we provide basic concepts in case retrieval. The need and process of case adaptation are explained briefly in Section 1.7. The issues of case learning and case-base maintenance are discussed in Section 1.8. In Section 1.9 an example is provided to demonstrate how a CBR system can be built. The question of whether CBR is a methodology or a technology is discussed in Section 1.10. Finally, in Section 1.11 the relevance to CBR of soft computing tools is explained.

1.2 COMPONENTS AND FEATURES OF CASE-BASED REASONING

Let us consider a medical diagnosis system as a typical example of using case-based reasoning in which the diagnosis of new patients is based on physicians' past experience. In this situation, a case could represent a person's symptoms together with the associated treatments. When faced with a new patient, the doctor compares the person's current symptoms with those of earlier patients who had similar symptoms. Treatment of those patients is then used and modified, if necessary, to suit the new patient (i.e., some adaptation of previous treatment may be needed). In real life there are many similar situations that employ the CBR paradigm to build reasoning systems, such as retrieving preceding law cases for legal arguments, determining house prices based on similar information from other real estate, forecasting weather conditions based on previous weather records, and synthesizing a material production schedule from previous plans.

From the examples above we see that a case-based reasoner solves new problems by adapting solutions to older problems. Therefore, CBR involves reasoning from prior examples: retaining a memory of previous problems and their solutions and solving new problems by reference to that knowledge. Generally, a case-based reasoner will be presented with a problem, either by a user or by a program or system. The case-based reasoner then searches its memory of past cases (called the *case base*) and attempts to find a case that has the same problem specification as the case under analysis. If the reasoner cannot find an identical case in its case base, it will attempt to find a case or multiple cases that most closely match the current case.

In situations where a previous identical case is retrieved, assuming that its solution was successful, it can be offered as a solution to the current problem. In the more likely situation that the case retrieved is not identical to the current case, an adaptation phase occurs. During adaptation, differences between the current and retrieved cases are first identified and then the solution associated with the case retrieved is modified, taking these differences into account. The solution returned in response to the current problem specification may then be tried in the appropriate domain setting.

The structure of a case-based reasoning system is usually devised in a manner that reflects separate stages: for example, for retrieval and adaptation, as described above. However, at the highest level of abstraction, a case-based reasoning system can be viewed as a black box (see Fig. 1.1) that incorporates the reasoning mechanism and the following external facets:

- The input specification or problem case
- The output that defines a suggested solution to the problem
- The memory of past cases, the case base, that are referenced by the reasoning mechanism

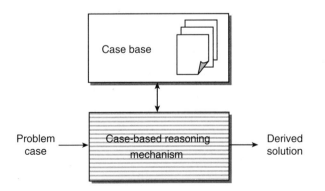

Figure 1.1   CBR system.

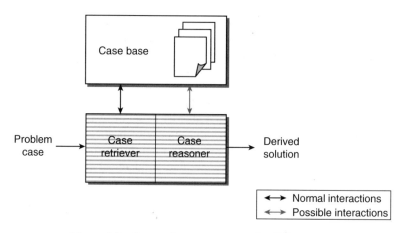

Figure 1.2   Two major components of a CBR system.

In most CBR systems, the *case-based reasoning mechanism*, alternatively referred to as the *problem solver* or *reasoner*, has an internal structure divided into two major parts: the case retriever and the case reasoner (see Fig. 1.2). The case retriever's task is to find the appropriate cases in the case base, while the case reasoner uses the cases retrieved to find a solution to the problem description given. This reasoning process generally involves both determining the differences between the cases retrieved and the current case, and modifying the solution to reflect these differences appropriately. The reasoning process may or may not involve retrieving additional cases or portions of cases from the case base.

1.2.1   CBR System versus Rule-Based System

The approach of case-based reasoning can be contrasted with that used in other knowledge-based systems, such as rule-based or combined frame-rule-based systems. In *rule-based systems*, one has a rule base consisting of a set of production rules of the form: IF A, THEN B, where A is a condition and B is an action. If the condition A holds true, the action B is carried out. Condition A can be a composite condition consisting of, say, a conjunction of premises $A_1, A_2, \ldots, A_n$. In addition, a rule-based system has an inference engine that compares the data it holds in working memory with the condition parts of rules to determine which rules to fire. *Combined frame-rule-based systems* also utilize frames, in addition to rules, to capture stereotypical knowledge. Frames consist of slots that can have default values, actual values, or attached daemons. Frames use a procedure or a rule set to determine the values required when they are triggered. Rule-based and combined frame-rule-based systems require one to acquire the symbolic knowledge that is represented in these rules or frames using manual knowledge engineering or automated knowledge acquisition tools. Sometimes, one utilizes a model of the problem as a basis for reasoning about a situation, where the model can be qualitative or quantitative. These systems are referred to as *model-based systems.*

Case-based reasoning systems are an alternative, in many situations, to rule-based systems. In many domains and processes, referring to cases as a means of reasoning can be an advantage due to the nature of this type of problem solving. One of the most time-consuming aspects when developing a rule-based system is the knowledge acquisition task. Acquiring domain-specific information and converting it into some formal representation can be a huge task and in some situations, especially those with less well understood domains, formalization of the knowledge cannot be done at all. Case-based systems usually require significantly less knowledge acquisition, since it involves collecting a set of past experiences without the added necessity of extracting a formal domain model from these cases. In many domains there are insufficient cases to extract a domain model, and this is another benefit of CBR: A system can be created with a small or limited amount of experience and then developed incrementally, adding more cases to the case base as they become available.

1.2.2   CBR versus Human Reasoning

The processes that make up case-based reasoning can be seen as a reflection of a particular type of human reasoning. In many situations, the problems that human beings encounter are solved with a human equivalent of CBR. When a person encounters a new situation or problem, he or she will often refer to a past experience of a similar problem. This previous experience may be one that they have had or one that another person has experienced. If the experience originates from another person, the case will have been added to the (human) reasoner's memory through either an oral or a written account of that experience.

In general, we have referred to case-based reasoning as being applied to problem solving. Case-based reasoning can also be used in other ways, most notably that of arguing a point of view. For example, many students will come to their teacher with various requests. A request might be for an extension to a deadline or for additional materials. It is a common experience of a teacher, after refusing one of these requests, to have a student argue the point. One of the common techniques that a student will use is to present evidence that in another course, or with another lecturer or teacher, their request has been granted in a similar situation, with similar underlying rules.

This sort of reasoning, very common in law domains, illustrates another way in which case-based reasoning systems can be implemented. Just as an attorney argues a point in court by references to previous cases and the precedents they set, CBR systems can refer to a case base containing court cases and find cases that have characteristics similar to those of the current one. The similarities may cover the entire case or only certain points that led to a portion of the ruling. Cases can therefore be discovered that may support some portions of the current case while opposing other parts. Case-based systems that perform this sort of argument are generally referred to as *interpretive reasoners.*

The idea of CBR is intuitively appealing because it is similar to human problem-solving behavior. People draw on past experience while solving new problems, and this approach is both convenient and effective, and it often relieves the burden of in-depth analysis of the problem domain. This leads to the advantage that CBR can be based on shallow knowledge and does not require significant effort in knowledge engineering when compared with other approaches (e.g., rule-based).

1.2.3 CBR Life Cycle

The problem-solving life cycle in a CBR system consists essentially of the following four parts (see Fig. 1.3):

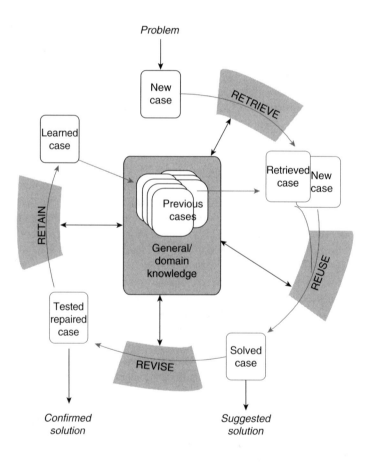

Figure 1.3   CBR cycle. (From [10].)

1. *Retrieving* similar previously experienced cases (e.g., problem–solution–outcome triples) whose problem is judged to be similar
2. *Reusing* the cases by copying or integrating the solutions from the cases retrieved
3. *Revising* or adapting the solution(s) retrieved in an attempt to solve the new problem
4. *Retaining* the new solution once it has been confirmed or validated

In many practical applications, the reuse and revise stages are sometimes difficult to distinguish, and several researchers use a single *adaptation* stage that replaces and combines them. However, adaptation in CBR systems is still an open question because it is a complicated process that tries to manipulate case solutions. Usually, this requires the development of a causal model between the problem space (i.e., the problem specification) and the solution space (i.e., the solution features) of the related cases.

In Figure 1.3, the cases stored in the case library (i.e., previous cases) were supplemented by general knowledge, which is usually domain dependent. This support may range from very weak to very strong, depending on the type of CBR method. For example, in using previous patient records for medical diagnosis, a causal model of pathology and anatomy may constitute the general knowledge used by a CBR system. This knowledge may be in the form of a set of IF–THEN rules or some preconditions in using the cases. Therefore, each stage in the CBR life cycle is associated with some tasks (see Fig. 1.4).

The process-oriented view of the CBR life cycle provides a global and external view of what is happening, while a task-oriented view is good for describing the actual mechanisms. Tasks are set up depending on the goals of the system, and a particular task is performed by applying one or more methods. In Figure 1.4, tasks are shown by node names, and the possible constituting methods appear in italic type. The links between task nodes (solid lines) represent various task decompositions. For example, the retrieval task is decomposed into the following tasks: identifying relevant descriptors, searching a set of past cases, matching the relevant descriptors to past cases, and selecting the most similar case(s). The methods under each task (dashed lines) indicate possible ways of completing the task. A method specifies an algorithm that identifies and controls execution of the particular subtask. The list of methods corresponding to a task shown in Figure 1.4, is not exhaustive. Selection of a suitable method depends on the problem at hand and requires knowledge of the application domain. In situations where information is incomplete or missing—and we want to exploit the tolerance for imprecision, uncertainty, approximate reasoning, and partial truth—soft computing techniques could provide solutions with tractability, robustness, and low cost.

Before we describe these issues and some of the operations (methods) under such major tasks as case representation and indexing, case retrieval, case adaptation, and case learning and case-base maintenance, in the following two sections we provide guidelines and advantages in the use of case-based reasoning.

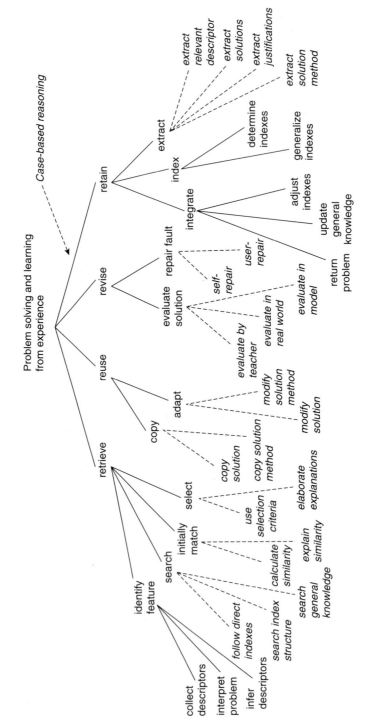

Figure 1.4 Task-method decomposition of CBR. (From [10].)

1.3 GUIDELINES FOR THE USE OF CASE-BASED REASONING

Although case-based reasoning is useful for various types of problems and domains, there are times when it is not the most appropriate methodology to employ. There are a number of characteristics of candidate problems and their domains, as mentioned below, that can be used to determine whether case-based reasoning is applicable [11–13]:

1. *Does the domain have an underlying model?* If the domain is impossible to understand completely or if the factors leading to the success or failure of a solution cannot be modeled explicitly (e.g., medical diagnosis or economic forecast), CBR allows us to work on past experience without a complete understanding of the underlying mechanism.

2. *Are there exceptions and novel cases?* Domains without novel or exceptional cases may be modeled better with rules, which could be determined inductively from past data. However, in a situation a where new experiences and exceptions are encountered frequently, it would be difficult to maintain consistency among the rules in the system. In that case the incremental case learning characteristics of CBR systems makes it a possible alternative to rule-based systems.

3. *Do cases recur?* If the experience of a case is not likely to be used for a new problem, because of a lack of similarity, there is little value in storing the case. In other words, when experiences are not similar enough to be compared and adapted successfully (i.e., cases do not recur), it might be better to build a model of the domain to derive the solution.

4. *Is there significant benefit in adapting past solutions?* One should consider whether there is a significant benefit, in terms of resources (e.g., system development time, processing effort), to creating a solution through modifying a similar solution rather than creating a solution to a problem from scratch.

5. *Are relevant previous cases obtainable?* Is it possible to obtain data that record the necessary characteristics of past cases? Do the recorded cases contain the relevant features of the problem and its context that influenced the outcome of the solution? Is the solution recorded in sufficient detail to allow it to be adapted in the future? These questions allow one to go for the CBR framework.

1.4 ADVANTAGES OF USING CASE-BASED REASONING

In this section we summarize some of the advantages of CBR from various points of view.

1. *Reducing the knowledge acquisition task.* By eliminating the need to extract of a model or a set of rules, as is necessary in model/rule-based systems, the

knowledge acquisition tasks of CBR consist primarily of the collection of relevant existing experiences/cases and their representation and storage.

2. *Avoiding repeating mistakes made in the past.* In systems that record failures as well as successes, and perhaps the reason for those failures, information about what caused failures in the past can be used to predict potential failures in the future.

3. *Providing flexibility in knowledge modeling.* Due to their rigidity in problem formulation and modeling, model-based systems sometimes cannot solve a problem that is on the boundary of their knowledge or scope or when there is missing or incomplete data. In contrast, case-based systems use past experience as the domain knowledge and can often provide a reasonable solution, through appropriate adaptation, to these types of problems.

4. *Reasoning in domains that have not been fully understood, defined, or modeled.* In a situation where insufficient knowledge exists to build a causal model of a domain or to derive a set of heuristics for it, a case-based reasoner can still be developed using only a small set of cases from the domain. The underlying theory of domain knowledge does not have to be quantified or understood entirely for a case-based reasoner to function.

5. *Making predictions of the probable success of a proffered solution.* When information is stored regarding the level of success of past solutions, the case-based reasoner may be able to predict the success of the solution suggested for a current problem. This is done by referring to the stored solutions, the level of success of these solutions, and the differences between the previous and current contexts of applying these solutions.

6. *Learning over time.* As CBR systems are used, they encounter more problem situations and create more solutions. If solution cases are tested subsequently in the real world and a level of success is determined for those solutions, these cases can be added to the case base and used to help in solving future problems. As cases are added, a CBR system should be able to reason in a wider variety of situations and with a higher degree of refinement and success.

7. *Reasoning in a domain with a small body of knowledge.* While in a problem domain for which only a few cases are available, a case-based reasoner can start with these few known cases and build its knowledge incrementally as cases are added. The addition of new cases will cause the system to expand in directions that are determined by the cases encountered in its problem-solving endeavors.

8. *Reasoning with incomplete or imprecise data and concepts.* As cases are retrieved, they may not be identical to the current case. Nevertheless, when they are within some defined measure of similarity to the present case, any incompleteness and imprecision can be dealt with by a case-based reasoner. Although these factors may cause a slight degradation in performance, due to the increased disparity between the current and retrieved cases, reasoning can continue.

9. *Avoiding repeating all the steps that need to be taken to arrive at a solution.* In problem domains that require significant processes to create a solution from scratch, the alternative approach of modifying an earlier solution can reduce this processing requirement significantly. In addition, reusing a previous solution also

allows the actual steps taken to reach that solution to be reused for solving other problems.

10. *Providing a means of explanation.* Case-based reasoning systems can supply a previous case and its (successful) solution to help convince a user of, or to justify, a proposed solution to the current problem. In most domains there will be occasions when a user wishes to be reassured about the quality of the solution provided by a system. By explaining how a previous case was successful in a situation, using the similarities between the cases and the reasoning involved in adaptation, a CBR system can explain its solution to a user. Even for a hybrid system, one that may be using multiple methods to find a solution, this proposed explanation mechanism can augment the causal (or other) explanation given to a user.

11. *Extending to many different purposes.* The number of ways in which a CBR system can be implemented is almost unlimited. It can be used for many purposes, such as creating a plan, making a diagnosis, and arguing a point of view. Therefore, the data dealt with by a CBR system are able to take many forms, and the retrieval and adaptation methods will also vary. Whenever stored past cases are being retrieved and adapted, case-based reasoning is said to be taking place.

12. *Extending to a broad range of domains.* As discussed in Chapter 6, CBR can be applied to extremely diverse application domains. This is due to the seemingly limitless number of ways of representing, indexing, retrieving, and adapting cases.

13. *Reflecting human reasoning.* As there are many situations where we, as humans, use a form of case-based reasoning, it is not difficult to convince implementers, users, and managers of the validity of the paradigm. Similarly, humans can understand a CBR system's reasoning and explanations and are able to be convinced of the validity of the solutions they receive from a system. If a human user is wary of the validity of an earlier solution, they are less likely to use this solution. The more critical the domain, the lower the chance that a past solution will be used and the greater the required level of a user's understanding and credulity.

We describe next, in brief, the four major tasks: case representation and indexing, case retrieval, case adaptation, and case learning and case-base maintenance.

1.5 CASE REPRESENTATION AND INDEXING

As mentioned earlier, a case can be said to be a record of a previous experience or problem. The information that is recorded about the experience will, by necessity, depend on the domain as well as the purpose for which this case will be used. For a problem-solving CBR system, the details will usually include specification of the problem and the relevant attributes of the environment that describe the circumstances of the problem. Another vital part of a case is a description of the solution that was used on an earlier occasion when a similar situation was encountered. Depending on how the CBR system reasons with cases, this solution may include only the facts that define a solution, or it may include information about additional steps or processes involved in obtaining a solution. It is also important to include a measure of success in the case description if the solutions (or cases) in the case base have achieved different levels of success or failure.

When a comparison is made between the knowledge stored in a model/rule-based system and that stored in a case base, it is apparent that the latter knowledge is of a more specific nature. While the knowledge in a model/rule-based system has been abstracted so that it is applicable in as wide a variety of situations as possible, the knowledge contained in a case-based system will remain specific to the case in which it is stored [13]. The specific knowledge of a case-based system means that related knowledge (i.e., knowledge applicable in a specific circumstance) is stored in close proximity. Thus, rather than drawing knowledge from a wider net, the knowledge needed to solve a specific problem can be found grouped together in a few or even one of the cases.

The case base in the CBR system is the memory of all cases stored previously. There are three general issues that have to be considered when creating a case base:

- The structure and representation of the cases
- The memory model used for organizing the entire case base
- The selection of indexes used to identify each case

1.5.1 Case Representation

Cases in a case base can represent many different types of knowledge that can be stored in many different representational formats. The intended purpose of a CBR system will greatly influence what is stored. For example, a case-based reasoning system may be aimed at the creation of a new design or plan, or the diagnosis of a new problem, or arguing a point of view using precedents. Therefore, in each type of CBR system, a case may represent something different. For example, the cases could represent people, objects, situations, diagnoses, designs, plans, or rulings, among many other representations. In many practical CBR applications, cases are usually represented as two unstructured sets of attribute–value pairs that represent the problem and solution features [14]. However, a decision as to exactly what to represent can be one of the most difficult decisions to make.

For example, in medical CBR systems performing patient diagnosis, a case could represent a person's entire medical history or be limited to a single visit to a doctor. In the latter situation, the case may be a set of symptoms together with a diagnosis. It may also include a prognosis or treatment. If a case represents a person, a more complete model is being used, since this could incorporate changing symptoms from one patient visit to the next. However, it is more difficult to find and use cases in the latter format, for example, when searching for a particular set of symptoms to obtain a diagnosis or treatment. Alternatively, if a case is simply a single visit to the doctor, involving only the symptoms at the time of that visit and a diagnosis of those symptoms, then changes in a patient's symptoms, which might be a useful key in solving a future problem, may be missed.

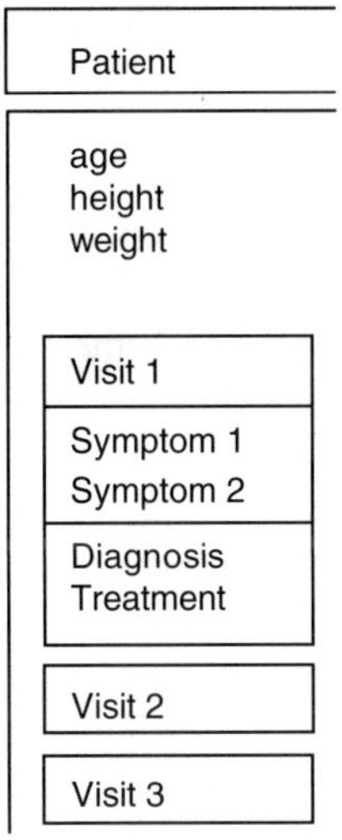

Figure 1.5 Patient case record.

In a situation such as the medical example described above, cases may need to be decomposed to their subcases. For example, a person's medical history could include as subcases all his or her visits to a doctor. In an object-oriented representation, this can be represented as shown in Figure 1.5.

Regardless of what a case actually represents as a whole, features have to be represented in some format. One of the advantages of case-based reasoning is the flexibility that this approach offers regarding representation. Depending on the types of features that have to be represented, an appropriate implementation platform can be chosen. This implementation platform ranges from simple Boolean, numeric, and textual data to binary files, time-dependent data, and relationships between data; CBR can be made to reason with any of these representation formats.

No matter how it is stored or the data format that is used to represent it, a case must store information that is both relevant to the purpose of the system and will also ensure that the most appropriate case is retrieved to solve each new problem situation. Thus, the cases have to include those features that will ensure that a case will be retrieved in the most appropriate context.

In many CBR systems, not all of the existing cases need to be stored. In these types of systems, specific criteria are needed to decide which cases will be stored and which will be discarded. For example, in a situation where two or more cases are very similar, only one case may need to be stored. Alternatively, it may be possible to create an artificial case that is a generalization of two or more cases that describe actual incidents or problems. By creating generalized cases, the most important aspects of a case need to be stored only once.

When choosing a representation format for a case, there are many choices and many factors to consider. Some examples of representation formats that may be used include database formats, frames, objects, and semantic networks. There are a number of factors that should be considered when choosing a representation format for a case:

- Segments within the cases (i.e., internal structure) that form natural subcases or components. The format chosen needs to be able to represent the various forms taken by this internal structure.
- Types and structures associated with the content or features that describe a case. These types have to be available, or be capable of being created, in the case representation.
- The language or shell chosen in which to implement the CBR system. The choice of a shell may limit the formats that can be used for representation. It should also be noted that the choice of language or shell is going to be influenced by a number of factors. The availability of various shells or languages, and the knowledge of the implementer, are the primary influences.
- The indexing and search mechanism planned. Cases have to be in a format that the case retrieval mechanism can deal with effectively.
- The form in which cases are available or obtained. For example, if a case base is to be formed from an existing collection of past experiences, the ease with which these experiences can be translated into an appropriate form for the CBR system could be important.

Whatever format is chosen to represent cases, the collection of cases itself also has to be structured in a way that facilitates retrieval of the appropriate case when queried. Numerous approaches have been used to index cases for efficient retrieval. A flat case base is a common structure. In this method indexes are chosen to represent the important aspects of the case, and retrieval involves comparing the query case's features to each case in the case base. Another common case-base structure is a hierarchical structure, which stores the cases by grouping them into appropriate categories to reduce the number of cases that have to be searched during a query. The memory model for a chosen form of case representation will depend on a number of factors:

- The representation used in the case base.
- The purpose of the CBR system. For example, a hierarchical structure is a natural choice for a system solving classification problems.
- The number and complexity of the cases being stored. As the number of cases grows in a case base, a structure that is searched sequentially will consume more time during retrieval (e.g., a flat case base).
- The number of features that are used for matching cases during a search.
- Whether some cases are similar enough to group together. Where cases fall into natural groupings, some structuring facility may be useful.
- How much is known about a specific domain. This influences the ability to determine whether cases are similar. For example, if little domain knowledge is available, case structuring is apt to be wrong.

In conclusion, cases are assumed to have two components: problem specification and solution. Normally, the problem specification consists of a set of attributes and values. The attributes of a case should define that case uniquely and should be sufficient to predict a solution for that case. The representation may be a simple flat data structure or a complex object hierarchy.

1.5.2   Case Indexing

Case indexing refers to assigning indexes to cases for future retrieval and comparison. The choice of indexes is important to enable retrieval of the right case at the right time. This is because the indexes of a case will determine in which context it will be retrieved in the future. There are some suggestions for choosing indexes in [13,15,16]. Indexes must be predictive in a useful manner. This means that indexes should reflect the important features of a case and the attributes that influence the outcome of the case, and describe the circumstances in which a case is expected to be retrieved in the future.

Indexes should be abstract enough to allow retrieval in all the circumstances in which a case will be useful, but not too abstract. When a case's indexes are too abstract, the case may be retrieved in too many situations or too much processing is required to match cases. Although assigning indexes is still largely a manual process and relies on human experts, various attempts at using automated methods have been proposed in the literature. For example, Bonzano et al. [17] use inductive techniques for learning local weights of features by comparing similar cases in a case base. This method can determine the features that are more important in predicting outcomes and improving retrieval. Bruninghaus and Ashley [18] employ a factor hierarchically (multilevel hierarchical knowledge that relates factors to normative concerns) in guiding machine learning programs to classify texts according to the factors and issues that apply. This method acts as an automatic filter removing irrelevant information. It structures the indexes into a factor hierarchy, which represents the kinds of circumstances that are important to users. Other indexing methods include indexing cases by features and by dimensions that are predictive across the entire problem domain [19], computing the differences between cases, adaptation guided indexing and retrieval [20], and explanation-based techniques.

1.6   CASE RETRIEVAL

Case retrieval is the process of finding, within a case base, those cases that are the closest to the current case. To carry out effective case retrieval, there must be selection criteria that determine how a case is judged to be appropriate for retrieval and a mechanism to control how the case base is searched. The selection criteria are necessary to determine which is the best case to retrieve, by determining how close the current case is to the cases stored.

The case selection criteria depend partly on what the case retriever is searching for in the case base. Most often the case retriever is searching for an entire case, the features of which are compared to those of the current case. However, there are times when only a portion of a case is being sought. This situation may arise because no full case exists and a solution is being synthesized by selecting portions of a number of cases. A similar situation is when a retrieved case is being modified by adopting a portion of another case in the case base.

The actual processes involved in retrieving a case from a case base is highly dependent on the memory model and indexing procedures used. Retrieval methods employed by researchers and implementers are extremely diverse, ranging from a simple nearest-neighbor search to the use of intelligent agents. We discuss both the most commonly used and traditional methods in the following sections.

Retrieval is a major research area in CBR. The most commonly investigated retrieval techniques, by far, are the k-nearest neighbors (k-NN), decision trees, and their derivatives. These techniques involve developing a similarity metric that allows closeness (i.e., similarity) among cases to be measured.

1. *Nearest-neighbor retrieval.* In nearest-neighbor retrieval, the case retrieved is chosen when the weighted sum of its features that match the current case is greater than other cases in the case base. In simple terms with all features weighted equally, a case that matches the present case on n features will be retrieved rather than a case that matches on only k features, where $k < n$. Features that are considered more important in a problem-solving situation may have their importance denoted by weighting them more heavily in the case-matching process.

2. *Inductive approaches.* When inductive approaches are used to determine the case-base structure, which determines the relative importance of features for discriminating among similar cases, the resulting hierarchical structure of the case base provides a reduced search space for the case retriever. This may, in turn, reduce the query search time.

3. *Knowledge-guided approaches.* Knowledge-guided approaches to retrieval use domain knowledge to determine the features of a case that are important for retrieving that case in the future. In some situations, different features of a case will have different levels of importance or contribution to the success levels associated with that case. As with inductive approaches to retrieval, knowledge-guided indexing may result in a hierarchical structure, which can be more effective for searching.

4. *Validated retrieval.* There have been numerous attempts at improving retrieval. One of these is validated retrieval, proposed by Simoudis [21], which consists of two phases. Phase 1 involves the retrieval of all cases that appear to be relevant to a problem, based on the main features of the present case. Phase 2 involves deriving more discriminating features from the initial group of retrieved cases to determine whether these cases are valid in the current situation. The advantage of validated retrieval is that inexpensive computational methods can be used to make the initial retrieval from the case base, while more expensive computational methods can be used in the second phase, where they are applied to only a subset of the case base.

There are a number of factors to consider when determining the method of retrieval:

- The number of cases to be searched
- The amount of domain knowledge available
- The ease of determining weightings for individual features
- Whether all cases should be indexed by the same features or whether each case may have features that vary in importance

Once a case has been retrieved, there is usually an analysis to determine whether that case is close enough to the problem case or whether the search parameters need to be modified and the search conducted again. If the right choice is made during this analysis, there can be a significant time saving. For example, the adaptation time required for a distant case could be significantly greater than searching again. When considering an analysis method for this decision, the following points should be considered:

- The time and resources required for adaptation
- The number of cases in the case base (i.e., how likely it is that there is a closer case)
- The time and resources required for search
- How much of the case base has already been searched

If we now review the processes involved in CBR that we have presented thus far, we can represent these succinctly, as shown in Figure 1.6.

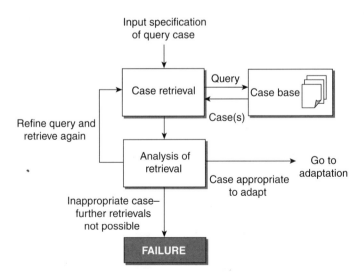

Figure 1.6 Processes involved in CBR.

1.7 CASE ADAPTATION

Case adaptation is the process of transforming a solution retrieved into a solution appropriate for the current problem. It has been argued that adaptation may be the most important step of CBR since it adds intelligence to what would otherwise be simple pattern matchers.

A number of approaches can be taken to carry out case adaptation:

- The solution returned (case retrieved) could be used as a solution to the current problem without modification, or with modifications where the solution is not entirely appropriate for the current situation.
- The steps or processes that were followed to obtain the earlier solution could be rerun without modification, or with modifications where the steps taken in the previous solution are not fully satisfactory in the current situation.
- Where more than one case has been retrieved, a solution could be derived from multiple cases or, alternatively, several alternative solutions could be presented.

Adaptation can use various techniques, including rules or further case-based reasoning on the finer-grained aspects of the case. When choosing a strategy for case adaptation, it may be helpful to consider the following:

- On average, how close will the case retrieved be to the present case?
- Generally, how many characteristics will differ between the cases?
- Are there commonsense or otherwise known rules that can be used when carrying out the adaptation?

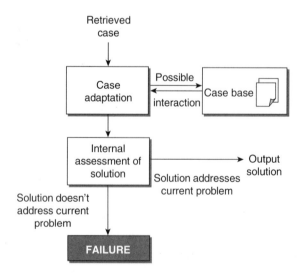

Figure 1.7 CBR entry into a learning state.

After the adaptation has been completed, it is desirable to check that the solution adapted takes into account the differences between the case retrieved and the current problem (i.e., whether the adaptation specifically addresses these differences). At this point, there is also a need to consider what action is to be taken if this check determines that the solution proposed is unlikely to be successful. At this stage, the solution developed is ready for testing an use in the applicable domain. This stage also concludes our description of all the necessary steps for any CBR system; however, many systems will now enter a learning phase, as shown in Figure 1.7.

1.8   CASE LEARNING AND CASE-BASE MAINTENANCE

1.8.1   Learning in CBR Systems

Once an appropriate solution has been generated and output, there is some expectation that the solution will be tested in reality (see Fig. 1.8). To test a solution, we have to consider both the way it may be tested and how the outcome of the test will be classified as a success or a failure. In other words, some criteria need to be defined for the performance rating of the proffered solution. Using this real-world assessment, a CBR system can be updated to take into account any new information uncovered in the processing of the new solution. This information can be added to a system for two purposes: first, the more information that is stored in a case base, the closer the match found in the case base is likely to be; second, adding information to the case base generally improves the solution that the system is able to create.

Learning may occur in a number of ways. The addition of a new problem, its solution, and the outcome to the case base is a common method. The addition of cases to the case base will increase the range of situations covered by the stored cases and reduce the average distance between an input vector and the closest stored vector (see Fig. 1.9).

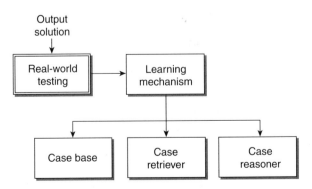

Figure 1.8   Learning mechanism in CBR.

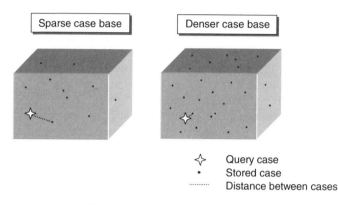

Figure 1.9   Distance between cases.

A second method of learning in a CBR system is using the solution's assessment to modify the indexes of the stored cases or to modify the criteria for case retrieval. If a case has indexes that are not relevant to the specific contexts in which it should be retrieved, adjusting the indexes may increase the correlation between the occasions when a case is actually retrieved and the occasions when it ought to have been retrieved. Similarly, assessment of a solution's performance may lead to an improved understanding of the underlying causal model of the domain that can be used to improve adaptation processing. If better ways can be found to modify cases with respect to the distance between the current and retrieved cases, the output solution will probably be improved.

When learning involves adding new cases to the case base, there are a number of considerations:

1. *In which situations should a case be added to the case base, and in which situations should it be discarded?* To determine an answer, we have to consider the level of success of the solution, how similar it is to other cases in the case base, and whether there are important lessons to be learned from the case.

2. *If a case is to be added to the case base, the indexes of the new case must be determined and how that case is to be added to the case base.* If the case base's structure and retrieval method are both highly structured (e.g., an inductively determined hierarchical structure or a set of neural networks), the incorporation of a new case may require significant planning and restructuring of the case base.

1.8.2   Case-Base Maintenance

When applying CBR systems for problem solving, there is always a trade-off between the number of cases to be stored in the case library and retrieval efficiency. The larger the case library, the greater the problem space covered; however, this would also downgrade system performance if the number of cases were to grow unacceptably high. Therefore, removing redundant or less useful cases to attain an acceptable error level is one of the most important tasks in maintaining CBR systems. Leake and Wilson [22] defined case-base maintenance as the implementation of policies for revising the organization or contents (representation, domain content, accounting information, or implementation) of a case base to facilitate future reasoning for a particular set of performance objectives.

The central idea of CBR maintenance is to develop some measures for case competence, which is the range of problems that a CBR system can solve. Various properties may be useful, such as the size, distribution, and density of cases in the case base; the coverage of individual cases; and the similarity and adaptation knowledge of a given system [23]. *Coverage* refers to the set of problems that each case could solve, and *reachability* refers to the set of cases that could provide solutions to the current problem [24]. The higher the density of cases, the greater the chances of having redundant cases. By expressing the density of cases as a function of case similarity, a suitable case deletion policy could be formulated for removing cases that are highly reachable from others.

Another reason for CBR maintenance is the possible existence of conflicting cases in the case library due to changes in domain knowledge or specific environments for a given task. For example, more powerful cases may exist that can contain inconsistent information, either with other parts of the same case or with original cases that are more primitive. Furthermore, if two cases are considered equivalent (with identical feature values), or if one case subsumes another by having more feature criteria, a maintenance process may be required to remove the redundant cases.

1.9 EXAMPLE OF BUILDING A CASE-BASED REASONING SYSTEM

So far, we have outlined the essential components and processes that make up a case-based reasoning system. For the convenience of readers, we provide here an example demonstrating how a CBR system can be built.

With the development of higher living standards and job opportunities in cities, more and more people move from various parts of a country to cities. When people reach a new place, they usually have to look for accommodations. In most cases, the main concern will be the renting of apartments. Some people may search through the World Wide Web (WWW) to look for rental information. However, case matching on the WWW is usually done in an exact manner. For example, if a person wants information on the rental of rooms or apartments in Lianhua Road, Pudong District, Shanghai City, it is very likely that no such information (no records) could be found, and therefore that search would fail. There are many reasons for such a failure; it may be because no apartments are available for rent, or the rental information is not available on the Web, or the search engine could not locate the information required. In this situation, CBR will be a suitable framework to provide answers to such queries because it is based on the degree of similarity

TABLE 1.1 Rental Information

Case	City District	Address of Apartment	Type of Apartment	Source of Information	Rental Cost of Apartment (yuan/month)
1	Mh	Lx	11	indi	500
2	Mh	Lz	21	indi	2,500
3	Mh	Ls1	21	indi	1,500
4	Mh	Xz	11	indi	500
5	Mh	Hm3	11	indi	880
6	Mh	Ls2	31	dn*	1,500
7	Mh	Dw	22	dn*	2,500
8	Xh	Cd	11	indi	1,200
9	Xh	Yd	32	indi	6,800
10	Xh	Xm	20	indi	1,600
11	Cn	Xh	11	indi	1,100
12	Cn	Fy	11	indi	1,100
13	Cn	Lj	21	dn*	3,500
14	Pd	Wl	21	indi	1,600
15	Pd	Wl	21	indi	1,600
16	Pd	Yg	10	indi	3,500
17	Pd	Ls	11	bc*	1,000
18	Yp	Xy	10	indi	750
19	Yp	Bx	32	indi	2,000
20	Yp	Cy	42	ys*	10,000

(not exactness) in matching with the capability of adaptation. Here we illustrate how to build a CBR system to handle this rental information searching problem.

Specifically, we explain how to perform case representation, case indexing, case similarity assessment, case adaptation, and case-base maintenance using a rental information searching problem. The example case base was taken from the Web site *http://sh.soufun.com/asp/rnl/leasecenter* on August 27, 2002, and consists of 251 cases. We randomly selected 20 cases (see Table 1.1) to constitute our sample. In the table, information is summarized. For example, "Mh" in the column "city district" represents Minghang; "11" in the column "type of apartment" represents one bedroom and one sitting room; and in the column "source of information," "indi" represents information obtained from individual landlords, and those marked by "*" represent information obtained from real estate agents.

The following steps are carried out using these 20 cases:

- *Case representation and indexing.* Choose an appropriate representation format for a case, and build a case index to facilitate the retrieval of similar cases in the future.
- *Case matching and retrieval.* Establish an appropriate similarity function, and retrieve cases that are similar to the present case.

TABLE 1.2   Case Representation

Case_id: 1	
Problem features	
City district	Mh
Address of apartment	Lx
Type of apartment	11
Source of information	indi
Solution	
Rental cost of apartment	500

- *Case adaptation.* Construct a suitable adaptation algorithm to obtain an appropriate solution for the query case.
- *Case-base maintenance.* Maintain consistency among cases (i.e., eliminate redundant cases) in the case base.

1.9.1   Case Representation

Representing cases is a challenging task in CBR. On the one hand, a case representation must be expressive enough for users to describe a case accurately. On the other hand, CBR systems must reason with cases in a computationally tractable fashion. Cases can be represented in a variety of forms, such as prepositional representations, frame representations, formlike representations, and combinations of these three. In this example we could choose a flat feature-value vector for the representation of cases as shown in Table 1.2.

1.9.2   Case Indexing

Table 1.2 describes the case representation. *Case indexing* is the method used to store these cases in computer memory. In this example we organize the cases hierarchically so that only a small subset of them needs to be searched during retrieval. Processes are as follows:

Step 1. Select the most important feature attribute, city district, as an index (in general, people think that the city district is the most important feature).

Step 2. Classify cases into five classes, $C_1, C_2, \ldots, C_5$ (Fig. 1.10) according to this feature.

When a new case *ne* is added to this index hierarchy, the following indexing procedure could be used. Classify (*ne*, CL) (i.e., classify *ne* as a member of one of the classes of CL), where CL $= \{C_1, C_2, \ldots, C_5\}$. Input a new case described by a set of features $ne = \{F_1, F_2, \ldots, F_5\}$.

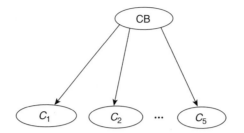

Figure 1.10   Indexing structure of the case base.

Step 1. Starting from $i = 1$, select one case e_j randomly from class C_i.
Step 2. Compare the value of the first feature (i.e., city district) of ne and e_j.
Step 3. If the two values are identical, $ne \in C_i$, else $i = i + 1$; go to step 1.
Step 4. Repeat steps 1, 2, and 3, until ne belongs to some class C_i.

1.9.3 Case Retrieval

Case retrieval is the process of finding within the case base the case(s) that are closest to the current case. For example, for a given case $\hat{e} = $ (Mh, Lz, 11, indi, θ), where we want to know the rent information θ, case retrieval is the process of finding those cases that are the closest to $\hat{e}$.

The retrieval algorithm is begun by deciding to which class the present case $\hat{e}$ belongs: Classify ($\hat{e}$, CL), where CL $= \{C_1, C_2, \ldots, C_5\}$ and $\hat{e}$ is a query case.

Step 1. Starting from $i = 1$, select one case e_j randomly from class C_i.
Step 2. Compare the feature value "district" of $\hat{e}$ and e_j.
Step 3. If the two values are identical, $\hat{e} \in C_i$, else $i = i + 1$; go to step 1.
Step 4. Repeat steps 1, 2, and 3 until $\hat{e}$ belongs to some class C_i.

Next, search for similar cases.

Step 1. If $\hat{e} \in C_i$, for each case e_i calculate the degree of similarity between $\hat{e}$ and e_i as

$$SM(\hat{e}, e_i) = \frac{\text{common}}{\text{common} + \text{different}}$$

where "common" represents the number of features whose value is the same between $\hat{e}$ and e_i, and "different" represents the number of features whose value is different between $\hat{e}$ and e_i.

Step 2. Rank the cases by similarity measure computed in step 1.

Step 3. Choose cases such as $e_1, e_2, \ldots, e_k$ from C_i which are most similar to $\hat{e}$; that is,

$$\text{SM}(\hat{e}, e_i) = \max\{\text{SM}(\hat{e}, e_j), e_j \in C_i\} \qquad i = 1, 2, \ldots, k$$

For the given query case $\hat{e} = (\text{Mh, Lz, 11, indi}, \theta)$, after following the algorithm above, four similar cases are retrieved from C_1: e_1, e_2, e_4, and e_5, where

$$e_1 = (\text{Mh, Lx, 11, indi, 500}) \qquad e_2 = (\text{Mh, Lz, 21, indi, 2500})$$
$$e_4 = (\text{Mh, Xz, 11, indi, 500}) \qquad e_5 = (\text{Mh, Hm3, 11, indi, 880})$$

and

$$\text{SM}(\hat{e}, e_1) = \text{SM}(\hat{e}, e_2) = \text{SM}(\hat{e}, e_4) = \text{SM}(\hat{e}, e_5) = 0.75$$

1.9.4 Case Adaptation

After a matching case is retrieved, a CBR system adapts the solution stored in the retrieved case to fulfill the needs of the current case. The following is an algorithm used to carry out the task of case adaptation.

Step 1. Group the retrieved cases (which are most similar to the present case $\hat{e}$) based on their solution value (i.e., rental cost of the apartment) into several categories, $C_1, C_2 \ldots, C_m$, with corresponding values of rental cost of the apartment, $a_1, a_2, \ldots, a_m$.

Step 2. Count the number of cases retrieved in each category $C_i (i = 1, 2, \ldots, m)$, and record them as $n_1, n_2, \ldots, n_m$.

Step 3. Choose category C_i for which $n_i = \max\{n_1, n_2 \ldots, n_m\}$.

Step 4. If there exists only one category C_i such that $n_i > n_j$, $j \neq i$, take the solution (rental cost of the apartment) stored in a case in C_i as the solution to the present case; else, if there exist several categories, such as $C_{i1}, C_{i2}, \ldots, C_{ik}$ satisfying $n_{i1} = n_{i2} = \cdots = n_{ik} = \max\{n_1, n_2, \ldots, n_m\}$, take the average value $\bar{b}$ defined by

$$\bar{b} = \frac{a_{i1} + a_{i2} + \cdots + a_{ik}}{k}$$

as the solution to the present case, where $a_{i1}, a_{i2}, \ldots, a_{ik}$ are values of the Rental cost of the apartment corresponding to $C_{i1}, C_{i2}, \ldots, C_{ik}$.

In Section 1.9.3 we retrieved four cases, e_1, e_2, e_4, e_5, which are most similar to the present case $\hat{e}$, where $\hat{e} = (\text{Mh, Lz, 11, indi}, \theta)$. According to the case adaptation algorithm above, we find a suggested (adapted) solution for this case. First, we group these four cases into three categories, $C_1 = \{e_1, e_4\}, C_2 = \{e_2\}$, and $C_3 = \{e_5\}$, where the corresponding solutions (the rental cost of the apartment stored in a case) are 500, 2500, and 880. We then count the number of cases in each category. For example, the number of cases in category C_1 is 2, the number of cases in category C_2 is 1, and the number of cases in category C_3 is 1. Since the number of cases in category C_1 is the greatest, we choose 500 as the suggested solution to the present case.

Note that once we have obtained an adapted solution, its validity should be tested in reality. In the example here, the landlord might be asked if the solution (i.e., the suggested price) is acceptable. If it is acceptable, the performance of this CBR system can then be improved by adding this new case record.

1.9.5 Case-Base Maintenance

The maintenance problem is very important, as it safeguards the stability and accuracy of CBR systems in real-world applications. However, our example here focuses only on the problem of case redundancy. From Table 1.1 we see that there exist some redundant cases, such as cases 14 and 15, that will downgrade system performance, so we should identify and remove these cases. A way of doing this is as follows:

Step 1. For each class C_i, as in Section 1.9.1, where we have the case base classified into five classes, classify all its cases by the address of the apartment and generate new classes $C_{i1}, C_{i2}, \ldots, C_{in}$ $(i = 1, 2, \ldots, 5)$.

Step 2. Count the number of cases in each class C_{ij}, $(i = 1, 2, \ldots, 5, j = 1, 2, \ldots, n)$.

Step 3. If the number of cases in C_{ij} is greater than or equal to 2, compare each feature value of these cases pairwise.

Step 4. If some cases have identical values for all features, retain one of these cases and remove the rest.

In this example, since the case library available is small, the principal task is to remove the redundant cases. However, with each increase in case library size, more maintenance will be required.

1.10 CASE-BASED REASONING: METHODOLOGY OR TECHNOLOGY?

Is CBR a technology, such as linear programming, neural networks, genetic algorithms, fuzzy logic, and probabilistic reasoning, or just a methodology for problem solving similar to structured systems analysis and design methodology? Janet Kolodner raised this question in 1993 [13]. She proposed the idea that CBR is both a cognitive model and a method of building intelligent systems. In 1999, Ian Watson published an article explicitly arguing that CBR is a methodology, not a technology [25]. In examining four very different CBR applications he showed that CBR describes a methodology for problem solving but does not pre-scribe specific technology. He pointed out that different techniques could be used and applied in various phases of the CBR problem-solving life cycle. For example, nearest-neighbor techniques, induction algorithms, fuzzy logic, and database techniques can all be applied to the retrieval phase of a CBR system (i.e., measuring similarity among cases). The equation

$$\text{similarity } (p, q) = \sum_{j=1}^{n} f(p_i, q_i) \times w_i \tag{1.1}$$

represents a typical nearest-neighbor technique that describes a situation for which p and q are two cases compared for similarity, n is the number of attributes in each case, i is an individual attribute from 1 to n, and w_i is the feature weight of attribute i. Generally, the similarity calculation continues until all cases in the case library have been compared, and ranked according to their similarity to a target problem case. Similarities are usually normalized to fall within the range 0 to 1, where 1 means a perfect match and 0 indicates a total mismatch. Methods other than nearest-neighbor techniques have also been used to determine similarity among cases (e.g., induction techniques such as ID3 and C4.5, fuzzy preference functions, and database techniques such as structured query language).

Since CBR is a methodology rather than a technology, inductions and many clustering algorithms, such as c-means clustering [26], Kohonen's self-organized network [27], and Fu's similarity matrix [28], could be used to partition a case library for similarity assessment. These techniques generally use three indexes as a measure of the clustering performance: intercluster similarity, intracluster similar-ity, and the total number of clusters. Many powerful commercial CBR tools have provided this function (e.g., Kate from AcknoSoft, ReCall from ISoft, CBR-Works from TecInno, and ReMind from Cognitive Systems) [29].

As suggested by Watson [25], if CBR is considered to be a methodology, this has a profound implication: CBR has no technology to call its own and therefore must use other technologies. It is this "CBR has no technology" scenario that creates a challenge for CBR researchers. Any number of technologies or approaches can be used and applied to the CBR cycle. Consequently, it is as a methodology that CBR's future is assured.

1.11   SOFT CASE-BASED REASONING

According to Lotfi Zadeh, soft computing is "an emerging approach to computing, which parallels the remarkable ability of the human mind to reason and learn in an environment of uncertainty and imprecision." In general, it is a consortium of computing tools and techniques, shared by closely related disciplines, including fuzzy logic (FL), neural network theory (NN), evolutionary computing (EC), and probabilistic reasoning (PR), with the latter discipline subsuming belief networks, chaos theory, and parts of learning theory. The development of rough set theory by Zdzislaw Pawlak adds an additional tool for dealing with vagueness and uncertainty. In soft computing, an individual tool may be used independently, depending on the application domains. Tools can also act synergistically, not competitively, to enhance an application domain of the other by integrating their individual merits (e.g., the uncertainty-handling capability of fuzzy sets, learning capability of artificial neural networks, and robust searching and optimization characteristics of genetic algorithms). The primary objective is to provide flexible information-processing systems that can exploit a tolerance for imprecision, uncertainty, approximate reasoning, and partial truth, in order to achieve tractability, robustness, low solution cost, and a closer resemblance to human decision making. By far the most successful hybrid systems are neuro-fuzzy systems [30], and it is anticipated that other types of hybrid systems, such as neuro-genetic, fuzzy-rough, neuro-fuzzy-genetic, and rough-neuro [31–36], will have a profound impact on the ways in which intelligent systems are designed and developed in the future.

The notion of *fuzzy sets* was introduced by Zadeh in 1965. It provides an approximate but effective and flexible way of representing, manipulating, and utilizing vaguely defined data and information. It can also describe the behaviors of systems that are too complex or too ill-defined to allow precise mathematical analysis using classical methods and tools. Unlike conventional sets, fuzzy sets include all elements of a universal set but with different membership values in the interval [0,1]. Similarly, in fuzzy logic, the assumption that a proposition is either true or false is extended to multiple-value logic, which can be interpreted as a degree of truth. The primary focus of fuzzy logic is on natural language, where it can provide a foundation for approximate reasoning using words (i.e., linguistic variables).

Artificial neural network models are attempts to emulate electronically the architecture and information representation scheme of biological neural networks. The collective computational abilities of densely interconnected nodes or processors may provide a natural technique in a manner analogous to human ability. *Neuro-fuzzy computing*, capturing the merits of fuzzy set theory and artificial neural networks, constitutes one of the best-known hybridizations in soft computing. This hybrid integration promises to provide more intelligent systems (in terms of parallelism, fault tolerance, adaptivity, and uncertainty management) able to handle real-life ambiguous recognition or decision-making problems.

Evolutionary computing (EC) involves adaptive techniques that are used to solve search and optimization problems inspired by the biological principles of natural selection and genetics. In EC, each individual is represented as a string of binary values; populations of competing individuals evolve over many generations according to some fitness function. A new generation is produced by selecting the best individuals and *mating* them to produce a new set of offspring. After many generations, the offspring contain all the most promising characteristics of a potential solution to the search problem.

Probabilistic computing has provided many useful techniques for the formalization of reasoning under uncertainty, in particular the Bayesian and belief functions and the Dempster–Shafer theory of evidence. The rough set approach deals mainly with the classification of data and synthesizing an approximation of particular concepts. It is also used to construct models that represent the underlying domain theory from a set of data. Often, in real-life situations, it is impossible to define a concept in a crisp manner. For example, given a specific object, it may not be possible to know to which particular class it belongs; the best knowledge derived from past experience may only give us enough information to conclude that this object belongs to a boundary between certain classes. Formulation of lower and upper set approximations can be generalized to some arbitrary level of precision which forms the basis for rough concept approximations.

As mentioned before, CBR is being recognized as an effective problem-solving methodology which constitutes a number of phases: case representation, indexing, similarity comparison, retrieval, and adaptation. For complicated real-world applications, some degree of fuzziness and uncertainty is almost always encountered. Soft computing techniques, such as fuzzy logic, neural networks, and genetic algorithms, will be very useful in areas where uncertainty, learning, or knowledge inference are part of a system's requirements. To gain an understanding of these techniques so as to identify their use in CBR, we summarize them briefly in the following sections. For the convenience of readers, a more detailed treatment of fuzzy logic, neural networks, genetic algorithms, and rough set theory are given in appendixes A to D.

1.11.1 Fuzzy Logic

Fuzzy set theory has been applied successfully to computing with words or the matching of linguistic terms for reasoning. In the context of CBR, using quantitative features to create indexes involves conversion of numerical features into qualitative terms for indexing and retrieval. These qualitative terms are always fuzzy. Moreover, one of the major issues in fuzzy set theory is measuring similarities in order to design robust systems. The notion of similarity measurement in CBR is also inherently fuzzy in nature. For example, Euclidean distances between features are always used to represent similarities among cases. However, the use of fuzzy set theory for indexing and retrieval has many advantages [37] over crisp measurements, such as the following:

- Numerical features could be converted to fuzzy terms to simplify comparison.
- Fuzzy sets allow multiple indexing of a case on a single feature with different degrees of membership.
- Fuzzy sets make it easier to transfer knowledge across domains.
- Fuzzy sets allow term modifiers to be used to increase flexibility in case retrieval.

Another application of fuzzy logic to CBR is the use of fuzzy production rules to guide case adaptations. For example, fuzzy production rules may be discovered by examining a case library and associating the similarity between problem and solution features of cases.

1.11.2   Neural Networks

Artificial neural networks (ANNs) are commonly used for learning and the generalization of knowledge and patterns. They are not appropriate for expert reasoning, and their abilities for explanation are extremely weak. Therefore, many applications of ANNs in CBR systems tend to employ a loosely integrated approach where the separate ANN components have specific objectives such as classification and pattern matching. Neural networks offer benefits when used for retrieving cases because case retrieval is essentially the matching of patterns: A current input case or pattern is matched with one or more stored cases or patterns. Neural networks are very good at matching patterns. They cope very well with incomplete data and imprecise inputs, which is of benefit in many domains, as some portion of the features is sometimes important for a new case, whereas other features are of little relevance. Domains that use case-based reasoning are usually complex. This means that the classification of cases at each level is normally nonlinear, and hence for each classification a single-layered network is not sufficient and a multilayered network is required.

Hybrid CBR and ANNs are a very common architecture for applications to solve complicated problems. Knowledge may first be extracted from the ANNs and represented by symbolic structures for later use by other CBR components. Alternatively, ANNs could be used for retrieval of cases where each output neuron represents one case.

1.11.3   Genetic Algorithms

Genetic algorithms (GAs) are adaptive techniques used to solve search and optimization problems, inspired by the biological principles of natural selection and genetics. In GA, each individual is represented as a string of binary values. Populations of competing individuals evolve over many generations according to some fitness function. A new generation is produced by selecting the best individuals and mating them to produce a new set of offspring. After many generations the offspring contain all the most promising characteristics of a potential solution to the search problem. Learning local and global weights of case features is one of the most popular applications of GAs to CBR. These weights indicate how important the features within a case are with respect to the solution features. Information about these weights can improve the design of retrieval methods and the accuracy of CBR systems.

1.11.4   Some CBR Tasks for Soft Computing Applications

As a summary, some of the tasks in the four major elements of the CBR cycle that have relevance as prospective candidates for soft applications are as follows:

- *Retrieve*: fuzzy indexing, connectionist indexing, fuzzy clustering and classification of cases, neural fuzzy techniques for similarity assessment, genetic algorithms for learning case similarity, probability and/or Bayesian models for case selection, case-based inference using fuzzy rules, fuzzy retrieval of cases, fuzzy feature weights learning, rough set–based methods for case retrieval

- *Reuse*: reusing cases by interactive and conversational fuzzy reasoning, learning reusable case knowledge, neural fuzzy approaches for case reuse
- *Revise*: adaptation of cases using neural networks and evolutionary approaches, mining adaptation rules using rough set theory, obtaining fuzzy adaptation knowledge from cases
- *Retain*: redundant case deletion using fuzzy rules, case reachability and coverage determination using neural networks and rough set theory, determination of case-base competence using fuzzy integrals

Although in this classification we have mentioned primarily the application of individual soft computing tools, various combinations can also be used [34].

1.12 SUMMARY

In this chapter we provide a brief explanation of case-based reasoning, its main components, its advantages, and the situations in which it is most useful. Next we outline briefly some of the most common soft computing techniques and their relevance to case-based reasoning. Then a simple example is used to explain the major ideas on which CBR systems are based. There are many important books and monographs that interested readers should consult for more information on these topics. A few suggestions for further reading follow.

- Special issue on soft case-based reasoning, *Applied Intelligence*, T. S. Dillon, S. K. Pal, and S. C. K. Shiu (eds.) (still to appear).
- *Soft Computing in Case-Based Reasoning*, S. K. Pal, T. S. Dillon, and D. S. Yeung (eds.), Springer-Verlag, London, 2001.
- *Applying Case-Based Reasoning: Techniques for Enterprise Systems*, I. Watson, Morgan Kaufmann, San Francisco, 1997.
- *Case-Based Reasoning Experiences, Lessons and Future Directions*, D. B. Leake (ed.), AAAI Press/MIT Press, Cambridge, MA, 1996.
- *Case-Based Reasoning*, J. Kolodner, Morgan Kaufmann, San Francisco, 1993.

In addition, the following Web sites provide much useful information as well as links to other CBR resources:

- D. Aha at the U.S. Naval Research Laboratory maintains the site *http://www.aic.nrl.navy.mil/~aha/*.
- R. Bergmann, I. Vollrath, and S. Schmitt at the University of Kaiserslautern maintain the site *http://www.cbr-web.org*.

- I. Watson at the University of Auckland in New Zealand maintains the site *http://www.ai-cbr.org*.

REFERENCES

1. R. Schank and R. Abelson (eds.), *Scripts, Plans, Goals and Understanding*, Lawrence Erlbaum Associates, Hillsdale, NJ, 1977.

2. R. Schank, *Dynamic Memory: A Theory of Reminding and Learning in Computers and People*, Cambridge University Press, New York, 1982.

3. J. L. Kolodner, Maintaining organization in a dynamic long term memory, *Cognitive Science*, vol. 7, pp. 243–280, 1983.

4. J. L. Kolodner, Reconstructive memory, a computer model, *Cognitive Science*, vol. 7, pp. 281–328, 1983.

5. R. L. Simpson, A computer model of case-based reasoning in problem solving: an investigation in the domain of dispute mediation, Ph.D. dissertation, School of Information and Computer Science, Georgia Institute of Technology, Atlanta, GA, 1985.

6. K. Sycara, Using case-based reasoning for plan adaptation and repair, in *Proceedings of the Case-Based Reasoning Workshop*, DARPA, Clearwater Beach, FL, Morgan Kaufmann, San Francisco, pp. 425–434, 1988.

7. K. J. Hammond, *Case-Based Planning*, Academic Press, San Diego, CA, 1989.

8. T. R. Hinrihs, *Problem Solving in Open Worlds*, Lawrence Erlbaum Associates, Hillsdale, NJ, 1992.

9. R. Bareiss, *Exemplar-Based Knowledge Acquisition: A Unified Approach to Concept Representation, Classification, and Learning*, Academic Press, San Diego, CA, 1989.

10. A. Aamodt and E. Plaza, Case-based reasoning: foundational issues, methodological variations, and system approach, *AI Communications*, vol. 7, no. 1, pp. 39–59, 1994.

11. J. L. Kolodner, An introduction to case-based reasoning, *Artificial Intelligence Review*, vol. 6, no. 1, pp. 3–34, 1992.

12. J. L. Kolodner and W. Mark, Case-based reasoning, *IEEE Expert*, vol. 7, no. 5, pp. 5–6, 1992.

13. J. L. Kolodner, *Case-Based Reasoning*, Morgan Kaufmann, San Francisco, pp. 27–28, 1993.

14. F. Gebhardt, A. Vob, W. Grather, and B. Schmidt-Beltz, *Reasoning with Complex Cases*, Kluwer Academic, Norwell, MA, 1997.

15. L. Birnbaum and G. Collins, Reminding and engineering design themes: a case study in indexing vocabulary, in *Proceedings of the DARPA Case-Based Reasoning Workshop*, San Francisco, Morgan Kaufmann, San Francisco, pp. 47–51, 1989.

16. K. J. Hammond, On functionally motivated vocabularies: an apologia, in *Proceedings of the Second DARPA Workshop on Case-Based Reasoning*, Pensacola Beach, FL, pp. 52–56, 1989.

17. A. Bonzano, P. Cunningham, and B. Smyth, Using introspective learning to improve retrieval in CBR: a case study in air traffic control, in *Proceedings of the Second International Conference on Case-Based Reasoning (ICCBR-97)*, Providence, RI, Springer-Verlag, Berlin, pp. 291–302, 1997.

18. S. Bruninghaus and K. D. Ashley, Using machine learning for assigning indices to textual cases, in *Proceedings of the Second International Conference on Case-Based Reasoning (ICCBR-97)*, Providence, RI, Springer-Verlag, Berlin, pp. 303–314, 1997.

19. T. Acorn and S. Walden, SMART: support management cultivated reasoning technology Compaq customer service, in *Proceedings of the Fourth Innovative Applications of Artificial Intelligence Conference on Artificial Intelligence (IAAI-92)*, San Jose, CA, AAAI Press, Menlo Park, CA, pp. 3–18, 1992.

20. B. Smyth and M. T. Keane, Adaptation-guided retrieval: questioning the similarity assumption in reasoning, *Artificial Intelligence*, vol. 102, pp. 249–293, 1998.

21. E. Simoudis, Using case-based retrieval for customer technical support, *IEEE Expert*, vol. 7, no. 5, pp. 7–12, 1992.

22. D. B. Leake and D. C. Wilson, Categorizing case-base maintenance: dimensions and directions, in *Proceedings of the Fourth European Workshop on Case-Based Reasoning (EWCBR-98)*, Dublin, Ireland, Springer-Verlag, Berlin, pp. 197–207, 1998.

23. B. Smyth and E. McKenna, Modeling the competence of case-bases, in *Proceedings of the Fourth European Workshop on Case-Based Reasoning (EWCBR-98)*, Dublin, Ireland, Springer-Verlag, Berlin, pp. 208–220, 1998.

24. K. Racine and Q. Yang, Maintaining unstructured case bases, in *Proceedings of the Second International Conference on Case-Based Reasoning (ICCBR-97)*, Providence, RI, Springer-Verlag, Berlin, pp. 553–564, 1997.

25. I. Watson, Case-based reasoning is a methodology, not a technology, *Knowledge-Based Systems*, vol. 12, pp. 303–308, 1999.

26. J. C. Bezdek, *Pattern Recognition with Fuzzy Objective Function Algorithms*, Plenum Press, New York, 1981.

27. T. Kohonen, *Self-Organization and Associate Memory*, Springer-Verlag, Berlin, 1988.

28. G. Fu, An algorithm for computing the transitive closure of a similarity matrix, *Fuzzy Sets and Systems*, vol. 51, pp. 189–194, 1992.

29. I. Watson, *Applying Case-Based Reasoning: Techniques for Enterprise Systems*, Morgan Kaufmann, San Francisco, 1997.

30. S. K. Pal and S. Mitra, *Neuro-Fuzzy Pattern Recognition: Methods in Soft Computing*, Wiley, New York, 1999.

31. S. K. Pal and P. P. Wang (eds.), *Genetic Algorithms for Pattern Recognition*, CRC Press, Boca Raton, FL, 1996.

32. S. K. Pal and A. Skowron (eds.), *Rough Fuzzy Hybridization: A New Trend in Decision Making*, Springer-Verlag, Singapore, 1999.

33. S. K. Pal, A. Ghosh, and M. K. Kundu (eds.), *Soft Computing for Image Processing*, Physica-Verlag, Heidelberg, 2000.

34. S. K. Pal, T. S. Dillon, and D. S. Yeung (eds.), *Soft Computing in Case-Based Reasoning*, Springer-Verlag, London, 2001.

35. S. K. Pal, W. Pedrycz, A. Skowron, and R. Świniarski, *Neurocomputing*, Vol. 36, Special volume on rough-neuro computing, Elsevier, Amsterdam, pp. 1–262, 2001.

36. S. K. Pal, L. Polkowski, and A. Skowron (eds.), *Rough-Neuro Approach: A Way for Computing with Words*, Springer-Verlag, Heidelberg, 2002.

37. B. C. Jeng and T. P. Liang, Fuzzy indexing and retrieval in case-based systems, *Expert Systems with Applications*, vol. 8, no. 1, pp. 135–142, 1995.

CHAPTER 2

CASE REPRESENTATION
AND INDEXING

2.1 INTRODUCTION

In general, cases can be considered as contextualized experiences, useful for reasoners to achieve their goals. Therefore, case representation can be viewed as the task of enabling the computer to recognize, store, and process past contextualized experiences. When considering case representation, we need to look at this problem from two points of view: first, the conceptual models that are used to design and represent cases, and second, the means of implementing cases in the computer (e.g., a relational table in a RDBMS; some data structures, such as B-trees and R-trees; or an object-oriented hierarchy). Selection of an appropriate scheme for case representation and indexing is essential because it provides the basic structure for which other CBR processes can be carried out.

As mentioned in Chapter 1, a case consists primarily of two parts: the problem part and the solution part. Each part can be further divided into smaller components that might be useful for later reasoning tasks, depending on the problem at hand. In this chapter we illustrate, how case representation and indexing can be done using traditional data modeling techniques, such as the relational database model, object-oriented model, and predicate logic model. Then the use of soft computing techniques for representing and indexing cases is described, such as the use of fuzzy sets, rough sets, and neural networks.

Perhaps the most important and commonly used traditional case representation technique is the *relational database approach*. A relation can be defined as a subset

Foundations of Soft Case-Based Reasoning. By Sankar K. Pal and Simon C. K. Shiu
ISBN 0-471-08635-5 Copyright © 2004 John Wiley & Sons, Inc.

of discrete objects of related domains. The relational database model is simple and flexible and has been adopted widely in many case-based reasoning applications. Each object (or case) is represented by a row in a relational table where the columns are used to define the attributes (or fields) of the objects. Hence, we could divide a relational table into two parts: a problem description part and a solution part. If the case has many relationships with other objects, or if the case can be broken down into subcases, a network of relationships can be developed. This relational database framework can reduce storage redundancy as well as improve retrieval efficiency. A simple example of buying a car is used to illustrate this idea later in the chapter.

Another popular traditional approach to case representation is the *object-oriented approach*. The advantage of this approach comes from its compact case representation ability, and the associated software reusability.

A third approach is the predicate logic–based technique. Mathematically speaking, a *predicate* denotes a relation among objects. However, in CBR, the term *predicate* is often used to represent a relationship between a production rule and some facts. Therefore, we could interpret that a case is, in fact, a collection of facts and predicates. The reasoning process in these CBR systems is carried out by firing the corresponding production rules contained in the case. One of the principal advantages of using predicates is that it allows the CBR system designer to incorporate many production rules in the system to form a hybrid rule/case-based reasoning system, which may be very effective in such application domains as fault diagnosis systems. However, the *predicate logic approach* has one major drawback: Retrieving data values from predicates for the purpose of comparing similarity among cases is not always as convenient as other approaches.

Although the traditional data models described above are useful to represent and to index cases, in many practical situations when specifying a case, it is often difficult to articulate the feature values precisely. This uncertainty may be caused by incomplete, missing, unquantifiable information, overlapping of the data regions, or user ignorance. Therefore, to make cases more expressive in dealing with such situations, soft computing techniques are introduced. Before we go into a detailed discussion of traditional and soft case representation and indexing methodologies, let us review briefly the concepts and ways of organizing cases in computer memory [1].

1. *Flat Memory, Serial Search.* Cases are stored sequentially in a simple list, array, or file. No explicit indexing structure is created, and searching is done sequentially, case by case, until the entire case library is searched. If the matching heuristic is effective, the best-matching case will always be retrieved. Adding new cases to a case library is also simple and straightforward (e.g., the case is added as the last element of a list).

2. *Shared-feature Networks, Breadth-first Graph Search.* Cases are organized hierarchically so that only a small subset of cases needs to be searched during retrieval. This subset must be likely to contain the best-matching or most useful cases. Shared-feature networks provide a means of clustering cases so that cases

with many shared features can be grouped together. Each of the internal nodes of a shared-feature network contains features shared by the cases below it, whereas the leaf nodes hold the actual cases. Building such a network is simple, and clustering algorithms are generally used to partition the cases. Individual features are used either to cluster or to differentiate the cases from others. Searching is done simply by matching the problem case against the contents of each node, starting at the highest level in the hierarchy. The best-matching node is chosen, and the same process is repeated among its descendant nodes until a case is returned with the closest match. There are many ways to organize cases into a shared-feature network, such as putting the most important features at the top of the hierarchy so that they are considered first. Many of the machine learning algorithms can be used to partition cases into different clusters. However, there are a number of disadvantages associated with organizing cases into such a network: for example, adding new cases will become a nontrivial task; it is difficult to keep the network optimal as cases are added; extra space is required for organization; to provide accurate retrieval for several reasoning goals, several shared-feature networks (each prioritized differently) might be needed; when an input is incomplete, no guidance is given on how to continue the search; and finally, there is no guarantee that good cases will not be missed.

3. *Discrimination Networks.* Each internal node represents a question that subdivides the set of cases stored in the hierarchy immediately beneath it. Hence, each child node represents a different answer to the question posed by its parent, and each child node organizes the cases that provide its answer. Discrimination networks make the search more efficient than with an unorganized list. They are also very efficient because this structure provides a more coherent indexing scheme where attributes are separated from values. However, when an input is incomplete, neither shared-feature networks nor discrimination networks give guidance on how to continue the search for a solution when a question in the network cannot be answered.

4. *Redundant Discrimination Networks.* A redundant discrimination network organizes cases using several different discrimination networks, each containing a different ordering of questions. These networks are searched in parallel.

5. *Flat Library, Parallel Search.* Using parallel processing machines to implement a flat case library may be one alternative for achieving both effectiveness and efficiency in retrieving and adding new cases to a case library. However, expensive hardware is needed, and it is difficult to implement complex matching functions.

6. *Hierarchical Memory, Parallel Search.* If the objective is to combine the advantages of a hierarchical memory with parallel retrieval algorithms, the strategy of hierarchical memory and parallel search is the option to choose. Although a parallel approach can make the algorithms execute faster, no matter how much parallelism is available, there will still be a need to partition the case library intelligently when it is becoming very large.

The remainder of the chapter is organized as follows. In Section 2.2 we describe some traditional methods of case representation, including relational database representation, object-oriented representation, and predicate representation. In Section 2.3 we provide case representation methodologies using fuzzy sets, rough sets, and rough-fuzzy hybridization, together with some experimental results. In Section 2.4 we explain both traditional and soft computing approaches for case indexing, including use of the Bayesian probability model, prototype-based incremental neural network, and three-layered back-propagation neural network.

2.2 TRADITIONAL METHODS OF CASE REPRESENTATION

As mentioned earlier, a case can be regarded as a contextualized piece of knowledge that could be structured in accordance with the real situation in the problem domain. Usually, in this structure there exists a goal describing the reasoner's aim and a set of constraints on this goal. Some information about the situation related to achieving the reasoner's goal may also be required. Another important component in this structure is the description of a solution, which may be some concepts or methods to achieve the goals. It may include the solution itself, the reasoning steps, the justification, the alternatives, and expectations regarding the outcome. Sometimes, an additional part called the *outcome part* can be added to a case, which includes feedback from the real world and interpretations of the feedback after carrying out the solutions. Different problem domains may have different case structuring schemes. For example, a cardiac diagnosing reasoner, Casey [2], does not have an outcome part. Casey collects information from patients and computes the similarity of the input data using the problem part, then selects the most closely matched case, and then adapts to give a diagnosis. The outcome part of cases is not necessary in Casey because the knowledge of Casey is already well defined and captured completely. However, in most real-world applications, it is difficult to obtain complete knowledge and safeguard it accuracy. Therefore, in this situation it is recommended that the outcome part be added to the case structure. This part records the status of the real world after the solution is carried out and is very useful for giving additional feedback to users.

To illustrate methods of representing cases, we use the following example of buying a car. To sell a car successfully, it is necessary for a salesperson to understand all the relevant car information that might be of interest to users (e.g., intended use, color, number of seats, and style). Other information relating to customers such as budget, preferences, and other special requirements, may also need to be considered. Figure 2.1 provides such a car-buying example. As stated earlier, this case consists of customer information, order information, supplier information, and automobile information. A typical case record is shown in Figure 2.2. Furthermore, a supplier produces many types of automobiles, which

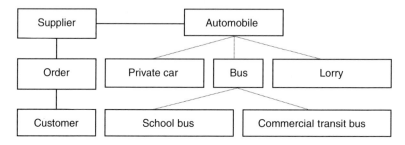

Figure 2.1   Car-buying example.

may be divided into three categories: private car, bus, and truck. There are two types of buses: school bus and commercial transit bus. In the following section we use the relational database model, object-oriented model, and the predicate logic model to represent this case example.

2.2.1   Relational Representation

Since we have nine objects (or entities) in the car-buying example, we need to create nine tables in our database, for one each object, to represent our cases. The tables are linked through the primary and foreign keys. The simplified SQL (structured query language) statements creating these relation tables is shown in Figure 2.3. (Attributes in bold type are selected as primary keys.) After creation of the tables, relation instances (i.e., cases) can be inserted into the table. Cases in the relational database tables can be retrieved later using SELECT-FROM-WHERE SQL statements.

An advantage of this relational database model is that it presents a multisided view of cases by normalizing the case data into third-normal form. However, this representation model has a drawback: Case-base designers must spend considerable time and energy to develop a relational database model for presenting a case.

Customer_Name:	Mike	Seats:	4
Customer_ID:	12345	Color:	Gray
Age:	30	Doors:	4
Country:	China	Liter:	2.5 L
Phone:	67890	Horsepower :	300
Profession:	Professor	Storage_Area:	20
Favorite_Color:	Silver	Load_Bearing:	10
		Fuel_Capacity:	18
Supplier_Phone:	54321	Date:	2003
Supplier_Name:	My Supplier		
		Recommendation:	My Supplier type 1
Price:	$ 55,000	Outcome:	Deal

Figure 2.2   Car-buying case record.

```
//CREATE DATABASE
CREATE TABLE
Automobile (Supplier_Phone, Auto_Type, Price, Seats, Doors, Color,
            Liter, Horsepower)


CREATE TABLE
PrivateCar (Auto_Type, Use, Entertainment_System, Airbag,
            Antitheft_System)


CREATE TABLE
Truck (Auto_Type, Storage_Area, Load_Bearing, Fuel_Capacity)


CREATE TABLE
Bus (Auto_Type, Carry_Passengers)


CREATE TABLE
SchoolBus (Auto_Type, Student_IDCheckSystem)


CREATE TABLE
CommercialTransitBus (Auto_Type, Entertainment_System,
                      Temperature_Control)


CREATE TABLE
Customer (Customer_Name, Customer_ID, Age, Country, Phone,
          Profession, Favorite_Color)


CREATE TABLE
Supplier (Supplier_Name, Supplier_Phone)


CREATE TABLE
Order(
     Order_ID,
     Customer_ID,
     Supplier_Phone,
     Auto_Type,
     Date,
     Recommendation,
     Outcome,
     )
```

Figure 2.3   Relational database representation of cases.

```
//INSERT VALUES INTO DATABASE
INSERT ("Honda","54321") INTO TABLE Supplier
...etc.


//FIND ORDER CASES IN WHICH A TRUCK HAS BEEN SOLD
SELECT  *
FROM    Customer, Order, Supplier, Automobile, Truck
WHERE   Customer.Customer_ID = Order.Customer_ID
AND     Supplier.Supplier_Phone = Order.Supplier_Phone
AND     Automobile.Auto_Type = Order.Auto_Type
AND     Truck.Auto_Type = Order.Auto_Type
```

Figure 2.3   (*Continued*)

2.2.2   Object-Oriented Representation

The popularity of object-oriented (OO) representation comes from its compact data-representing ability and associated software reusability. For case-based reasoning systems in CAD/CAM, multimedia, and global information systems, the standard relational database model is not suitable for building a complex case data structure. In this situation, OO representation works much better. In Figure 2.4

```
CREATE TABLE
Automobile (Supplier_Phone, Price, Seats, Doors, Color,
            Liter, Horsepower)


CREATE TABLE
PrivateCar (Auto_Type, Use, Entertainment_System, Airbag,
            Antitheft_System)
UNDER Automobile

CREATE TABLE
Truck (Auto_Type, Storage_Area, Load_Bearing, Fuel_Capacity)
UNDER Automobile

CREATE TABLE
Bus (Carry_Passengers)
UNDER Automobile

CREATE TABLE
SchoolBus (Auto_Type, Student_IDCheckSystem)
UNDER Bus
```

Figure 2.4   Object-oriented case representation.

```
CREATE TABLE
CommercialTransitBus (Auto_Type, Entertainment_System,
                      Temperature_Control)
UNDER Bus

CREATE TABLE
Supplier (Supplier_Name, Supplier_Phone)

CREATE TABLE
 Customer (Customer_Name, Customer_ID, Age, Country, Phone,
           Profession, Favorite_Color)

CREATE TABLE
Order(
      Order_ID,
      Cust REF(Customer), Supp REF(Supplier),
      Priv  REF(PrivateCar), Lorr REF(Truck),
      Scho REF(SchoolBus), Comtrans REF(CommercialTransitBus),
      Recommendation ,
      Outcome,
      Date)
//INSER VALULES INTO DATABASE
//FIND ORDER CASES
SELECT  *
FROM  Order
```

Figure 2.4 (*Continued*)

we use xSQL (extended structured query language)-style statements to represent the car-buying example. As can be seen, the OO method needs less memory storage to represent each case. Furthermore, since OO is a natural way of representing IS-A and HAS-A relationships, case representation is easier for users to understand.

2.2.3 Predicate Representation

A predicate is a relation among objects, and it consists of a condition part and an action part, IF (condition) and THEN (action). Predicates that have no conditional part are facts. Cases can be represented as a collection of predicates. The advantage of predicate representation is that it uses both rules and facts to represent a case, and it enables a case-base designer to build hybrid systems that are integrated rule/case-based. The Prolog clauses shown in Figure 2.5 represent the car-buying example.

```
FACTS
Order(
Customer (Customer_Name, Customer_ID, Age, Country, Phone,
          Profession, FavoriteColor),
Supplier (Supplier_Name, Supplier_Phone),

PrivateCar (Supplier_Phone, Auto_Type, Price, Seats, Doors, Color,
            Liter, Horsepower, Use, EntertainmentSystem,
            Airbag, AntitheftSystem),

SchoolBus (Supplier_Phone, Auto_Type, Price, Seats, Doors, Color,
           Liter, Horsepower, Student_IDCheckSystem),

CommercialTransitBus (Supplier_Phone, Auto_Type, Price, Seats, Doors,
                      Color, Liter, Horsepower, EntertainmentSystem,
                      Temperature_Control),

Truck (SupplierPhone, AutoType, Producer, Price, Seats, Doors, Color,
       Liter, Horsepower, StorageArea, LoadBearing, FuelCapacity),

Date,
Recommendation ,
Outcome
)

//INSERT FACTS
//QUERY CASES
Order (Customer(CuVariable1, CuVariable2, …, CuVariable8),
       Supplier (SuVariable1, SuVariable2),
       PrivateCar (Pvariable1, Pvariable2, …, Pvariable29),
       SchoolBus (ScVariable1, ScVariable2, …, ScVariable22),
       CommercialTransitBus (CoVariable1, Co Variable2, …,
                             CoVariable23),
       Truck (Tvariable1, Tvariable2, …, Tvariable24),
       Recommendation3,
       Date,
       OrderID)
```

Figure 2.5   Predicate schema of case representation.

2.2.4   Comparison of Case Representations

A comparison of the three case representation schemes above is shown in Table 2.1. The OO approach offers the highest compactness in terms of storage space and supports software reusability. The predicate logic approach is suitable only for a

TABLE 2.1   Comparison of Traditional Case Representations

	Relational Approach	Object-Oriented Approach	Predicate Logic Approach
Compactness	Medium	High	Low
Application independency	Yes	No	No
Software reusability	No	Yes	No
Case-base scale	Large	Large	Small
Retrieving feature values for computation	Easy	Easy	Difficult
Compatibility with rule-based system	Poor	Poor	Good and suitable for NLP
Case organization method	Keys	Inheritance/ reference	Data definition

small-scale case base and would become unmanageable if the number of cases grew too large. Furthermore, retrieving feature values for computation is difficult using the predicate logic approach. However, it does offer the opportunity to integrate with rule-based systems. Recently, XML (extensible markup language) has emerged as the standard for data representation and exchange on the Web. It is platform independent, object oriented, and can be well interfaced to many programming environments. XML is therefore expected to become a suitable tool in building open CBR systems on the Web in the near future.

2.3 SOFT COMPUTING TECHNIQUES FOR CASE REPRESENTATION

In this section we describe how the concept of fuzzy sets can be used to represent and index cases. This is followed by a method demonstrating an integration of rough and fuzzy sets for case generation and indexing.

2.3.1 Case Knowledge Representation Based on Fuzzy Sets

A fuzzy set A is a collection of objects drawn from the universal set U, with a continuum of grades of membership where each object x ($x \in U$) is assigned a membership value that represents the degree to which x fits the imprecise concept represented by the set A. Formally, it is written as follows:

$$A = \{\mu_A(x)/x, x \in U\} \qquad (2.1)$$

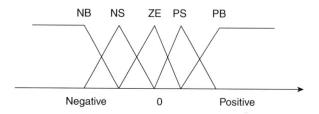

Figure 2.6 Fuzzy membership functions.

where $\mu_A(x)$, the *membership function*, is defined as

$$\mu_A: U \rightarrow [0, 1] \qquad (2.2)$$

The basic concepts and definitions of fuzzy set theory are described in Appendix A.

A *linguistic term* is a natural language expression or word that quantifies imprecisely an attribute such as income, age, height, and safety. Linguistic terms of this kind can be labels for fuzzy sets. In a typical application, the number of linguistic terms for each attribute in a case can be assumed to be five, usually referred to as negative big, negative small, zero, positive small, and positive big, or NB, NS, ZE, PS, and PB. Their membership functions can be expressed in many forms, such as in trapezoidal, Gaussian, and generalized bell shapes. So far, the most commonly used membership functions are triangular in shape, as shown in Figure 2.6. A fuzzy number, used to handle inexactly specified quantities, is usually formulated by a linguistic expression reflecting the closeness to a real number, for example, "around 300" or "more or less 1 million." Fuzzy numbers are fuzzy sets defined using equations (2.1) and (2.2).

2.3.1.1 Fuzzy Features Refer to the earlier car-buying example. The case features could be fuzzified into fuzzy linguistic terms and fuzzy numbers. The attribute that is suitable to be fuzzified is the price of a car. Usually, the car price fluctuates and varies with different models, colors, and components, so sometimes it is difficult for the user to specify the exact price of a car. In this situation, a fuzzy CBR system could accept a query such as "I am looking for a comfortable car that is moderately expensive and very durable." The architecture of the fuzzy CBR system is shown in Figure 2.7. The fuzzifier consists of a set of fuzzifying (and defuzzifying) algorithms that are used to generate fuzzy inputs from crisp user inputs. The fuzzy CBR system consists of the fuzzy similarity function for comparing cases in the case base with the input case. In the sample car record shown in Figure 2.2, the fuzzy sets of the price attribute can take the form of bell-shaped functions: very low, low, middle, high, and very high (see Fig. 2.8).

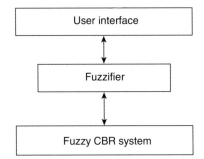

Figure 2.7 Fuzzy CBR for car buying.

The following equation shows the typical bell-shaped membership function of "very low":

$$\mu_{\text{very low}}(x) = \begin{cases} 1 & 0 \leq x \leq 20 \\ \left[1 + \left(\dfrac{x - 20}{8}\right)^4\right]^{-1} & 20 \leq x \leq 100 \end{cases} \qquad (2.3)$$

Therefore, given a price value, we could calculate its membership value to each fuzzy set. The fuzzy linguistic term with maximum value can then be chosen as the fuzzy input. For example, the membership value of price $55,000 in each category is as follows:

$$\mu_{\text{very low}}(55,000) = 0.0027$$
$$\mu_{\text{low}}(55,000) = 0.025$$
$$\mu_{\text{middle}}(55,000) = 0.8681$$
$$\mu_{\text{high}}(55,000) = 0.2907$$
$$\mu_{\text{very high}}(55,000) = 0.0104$$

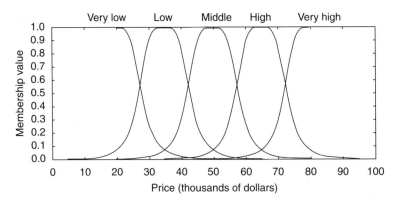

Figure 2.8 Bell-shaped fuzzy membership functions of car price.

Since the price \$55,000 has the highest membership value in the fuzzy set "middle," we could represent the price of the car as medium. Therefore, alternatively one can represent the attribute "price of the car" as a fuzzy set "middle" with the corresponding membership value.

2.3.1.2 Example The use of fuzzy set theory allows flexible encoding of case characteristics as real numbers, linguistic terms, fuzzy numbers, and fuzzy complex objects. These case features can also be organized systematically as fuzzy categories (i.e., clusters) or fuzzy granules. As we mentioned before, traditional methods of case knowledge representation are usually rigid. However, in many practical situations, when specifying a query case, it is often difficult to articulate the feature values precisely. If the cases can be organized in conceptually overlapping categories, retrieval could be implemented via a classification and similarity regime which encodes feature characteristics in a mixed fuzzy and crisp format. Use of the fuzzy concept significantly improves the flexibility and expressiveness with which case knowledge can be represented.

In Dubitzky et al. [3], a fuzzy feature representation method is applied to design a coronary heart disease (CHD) medical system in Northern Ireland in which the risk type is fuzzified into three categories: C_I, C_{II} and C_{III}. Category C_I corresponds to subjects that have a high CHD risk because of their cholesterol levels, cases C_{III} constitute stress-related risk types and cases in category C_{II} incline to risk that is caused by both factors. A target case e is described by the following combination of feature instances: $e: = \ll$ "almost never," "occasionally" $>$, $<8.00, 3.70\gg$; where the first pair represents the stress ([nervousness, worry]), and the second the cholesterol ([total, LDL]) values. To classify the target case e, its constituent feature values need to be matched against the fuzzy membership functions (f-functions) of the corresponding scales in the three categories (see Fig. 2.9). For example, if the feature of the target case (total $= 8.00$) yields membership values such as $f_1 = 0.16$ (in C_I) and f_2 and $f_3 = 0$ (in C_{II} and C_{III}), it is likely that the target case is a member of the high-cholesterol, low-stress category C_I.

2.3.2 Rough Sets and Determining Reducts

The theory of rough sets provides a new mathematical framework for analysis of data that are imprecise, vague, and uncertain. The main notion of rough sets is that if objects are similar or indiscernible in a given information system, the same set of information can be used to characterize them. Therefore, there may exist some objects that cannot be classified with certainty as members of the set or of its complement. In a rough set–theoretic approach, each vague concept is characterized by a pair of precise concepts (called the *lower* and *upper approximations* of the vague concept). Rough set approaches [4–6] deal primarily with the classification of data and synthesizing approximations of concepts. These approaches are also used to construct models that represent the underlying domain theory from a set of data. In CBR system development, it is often necessary to determine some properties

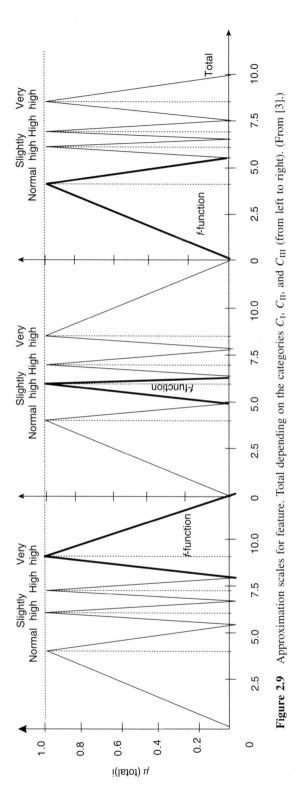

Figure 2.9 Approximation scales for feature. Total depending on the categories C_I, C_{II}, and C_{III} (from left to right). (From [3].)

TABLE 2.2   Patient Table

Patient	Condition C_1	Condition C_2	Condition C_3	Illness
1	High	Fair	Medium	θ_1
2	Very high	Good	Medium	θ_1
3	High	Good	Low	θ_1
4	Medium	Fair	Medium	θ_2
5	Very high	Fair	Low	θ_1
6	High	Good	Low	θ_2

from the case data to partition the cases into subsets. However, in real-life situations, it is sometimes impossible to define a classifying concept in a precise (crisp) manner. For example, given a new case, it may not be possible to know to which class it belongs. The best knowledge derived from past cases may only give us enough information to say that this new case belongs to a boundary between certain cases, which may consist of various possible solutions. The formulation of these lower- and upper-set approximations can be generalized to some arbitrary level of precision which forms the basis of rough concept approximations.

Some basic definitions and examples of rough set theory are described in Appendix D. Let us now look at an example (see Table 2.2) that illustrates the concept of rough sets. Let us consider a patient information system consisting of six patients, where each one is described by three health conditions: C_1, C_2, and C_3, and an illness by θ_1 or θ_2. Now the question is how to describe an illness uniquely (i.e., θ_1 or θ_2) in terms of the health conditions. After analyzing the data, it is clear that this question cannot be answered uniquely. For example, patients 3 and 6 have the same conditions, but patient 3 suffers from θ_1, whereas patient 6 suffers from θ_2. Hence, it is impossible to give a unique description of illness θ_1 or θ_2. The best we could decide is that patients 1, 2, and 5 surely suffer from illness θ_1, whereas patients 1, 2, 3, 5, and 6 possibly suffer from illness θ_1. Similarly, patient 4 surely suffers from illness θ_2, whereas patients 3, 4, and 6 possibly suffer from illness θ_2. Since it is impossible to give the unique characteristics of θ_1 or θ_2, a rough set–based technique can be used. Two sets corresponding to the concepts of "surely" and "possibly," the lower and upper approximations for each illness, respectively, describe the characteristics of θ_1 and θ_2.

The formal definition of rough sets given by Pawlak [4] is as follows: Any subset B of A determines a binary relation I_B on U, called an *indiscernibility relation*, and defined as $x\,I_B\,y$ if and only if $\text{Attr}(x) = \text{Attr}(y)$ for every $\text{Attr} \in B$, where $\text{Attr}(x)$ denotes the value of attribute a for element x. I_B is obviously an equivalence relation. The family of all equivalence classes of I_B (i.e., the partition determined by B) will be denoted by U/I_B, or simply U/B; an equivalence class of I_B (i.e., the block of the partition U/B) containing x will be denoted by $[x]_B$. If (x,y) belongs to I_B, x and y will be considered as B-indiscernible. Equivalence classes of the relation I_B or blocks of the partition U/B are referred to as *B-elementary concepts* of *B*-granules. The two operations defined on set X will be called the *B-lower* and the *B-upper approximation* of X (see Appendix D):

$$\underline{B}X = \{x \in U : [x]_B \subseteq X\} \tag{2.4}$$

$$\overline{B}X = \{x \in U : [x]_B \cap X \neq \emptyset\} \tag{2.5}$$

The following set will be referred to as the *B-boundary region* of X:

$$BN_B X = \overline{B}X - \underline{B}X \tag{2.6}$$

If the boundary region of X is not the empty set (i.e., $BN_B X \neq \emptyset$), the set X is referred to as *rough* (*inexact*) with respect to B. Rough sets can be defined using a rough membership function:

$$\mu_X^B(x) = \frac{|X \cap [x]_B|}{|[x]_B|} \quad \text{and} \quad \mu_X^B(x) \in [0,1] \tag{2.7}$$

The value of the membership function $\mu_X(x)$ represents a kind of conditional probability, and it can be interpreted as a degree of certainty to which x belongs to X [or $1 - \mu_X(x)$ as a degree of uncertainty].

Rough set theory can be applied to determine reducts on the decision table. The reduction process is performed on the positive region of the data set that is identified with two equivalence relations: the condition relation and the decision relation. The output is a simplified table that can work more efficiently than the original one. The related concepts of the positive region, the D-dispensable and the reduct, are described below.

- *Positive region.* Let C and D be two equivalence relations over U. The C-positive region of D, denoted as $\text{POS}_C(D)$, is defined as

$$\text{POS}_C(D) = \bigcup_{X \in U/D} \underline{C}X \tag{2.8}$$

- *D-dispensable.* $c \in C$ is D-dispensable in C if

$$\text{POS}_C(D) = \text{POS}_{C-\{c\}}(D) \tag{2.9}$$

Otherwise, c is D-indispensable in C.
- *D-reduct.* If every $c \in C$ is D-indispensable, then C is D-independent. The attributes set $R \subseteq C$ will be called a *D-reduct* of C if and only if R is D-independent and $\text{POS}_R(D) = \text{POS}_C(D)$.

TABLE 2.3   Car-Buying Cases

Case	Price (a)	Horsepower (b)	User Age (c)	Recommendation (d)
1	High	Medium	Old	Car type 1
2	Low	Low	Young	Car type 1
3	High	Medium	Young	Car type 2
4	Medium	Medium	Old	Car type 2
5	Medium	High	Old	Car type 3
6	High	Medium	Old	Car type 1
7	High	Medium	Young	Car type 2
8	Medium	Medium	Old	Car type 2
9	Medium	High	Old	Car type 3
10	Low	Low	Young	Car type 3

To illustrate the reduction process, the car-buying example could be represented as shown in Table 2.3. In the table the attributes "price," "horsepower," and "user age" comprise the conditional attributes set, denoted $C = \{a, b, c\}$, and "recommendation" comprises the decision part $D = \{d\}$.

The steps in determining reducts on the equivalence classes are as follows:

Step 1. Make partitions on U with decision relation D. The recommendation attribute has three values: car type 1, car type 2, and car type 3. Each value generates a partition on U. For instance, cases 1, 2, and 6 are of the same car type, so they belong to the same partition, X_1. Similarly, cases 3, 4, 7, and 8 belong to partition X_2, and cases 5, 9, and 10 belong to X_3.

$$X_1 = \{1,2,6\}, \qquad X_2 = \{3,4,7,8\}, \qquad X_3 = \{5,9,10\}$$

Step 2. For each partition X_i, identify the C-lower approximation of X_i, $i = 1, 2, 3$. For example, to identify the C-lower approximation of $X_1 = \{1,2,6\}$, the C equivalence class of each point in X_1 is checked:

$$[1]_c = \{1,6\} \subset X_1, \qquad [2]_c = \{2,10\} \not\subset X_1, \qquad [6]_c = \{1,6\} \subset X_1$$

Since case 2 does not satisfy the definition of C-lower approximation of X_1, it is removed from $\underline{C}X_1$; thus, $\underline{C}X_1 = \{1,6\}$. Similarly, $\underline{C}X_2 = \{3,4,7,8\}$ and $\underline{C}X_3 = \{5,9\}$.

Step 3. Build the positive region by combining all the C-lower approximations of the partitions:

$$\text{POS}_C(D) = \{1,3,4,5,6,7,8,9\}$$

TABLE 2.4   C-Equivalence Classes in the Positive Region

Class	Price (a)	Horsepower (b)	User Age (c)	Recommendation (d)	Number of Cases
Equiv_1	High	Medium	Old	Car type 1	2
Equiv_2	High	Medium	Young	Car type 2	2
Equiv_3	Medium	Medium	Old	Car type 2	2
Equiv_4	Medium	High	Old	Car type 3	2

Step 4. In the positive region, merge the cases with equivalent attribute values of relation C to form the equivalence classes. There are a total of four equivalence classes in $POS_C(D)$:

$$\text{Equiv\_1} = \{1,6\} \qquad \text{Equiv\_2} = \{3,7\} \qquad \text{Equiv\_3} = \{4,8\} \qquad \text{Equiv\_4} = \{5,9\}$$

These equivalence classes are shown in Table 2.4.

Step 5. Build the discerning matrix $\text{Discern} = (dis_{ij})_{4\times4}$ to get the reduct attributes set, where

$$dis_{ij} = \{r | r \in C, r(\text{Equiv\_}i) \neq r(\text{Equiv\_}j)\} \tag{2.10}$$

and $r(\text{Equiv\_}i)$ is the value of attribute r in equivalence class Equiv_i. For example, in Table 2.4 the entry that can discern Equiv_1 and Equiv_2 is attribute c (i.e., user age). Using this procedure, we could get all the entries of the discerning matrix (see Table 2.5).

Reduct_i of an equivalence class Equiv_i should be able to distinguish Equiv_i from all other equivalence classes. Therefore, Reduct_i should be the joint of the entries in the ith row of the discerning matrix. So computation of the reduct can be translated to computation of a Boolean function. For example, Reduct_2 is computed as

$$\begin{aligned} \text{Reduct\_2} &= c \wedge (a \vee c) \wedge (a \vee b \vee c) \\ &= c \wedge (a \vee b \vee c) \\ &= c \end{aligned}$$

Similarly, Reduct_1 $= a \wedge c$, Reduct_3 $= a \wedge b$, and Reduct_4 $= b$.

TABLE 2.5   Discerning Matrix of the Equivalence Classes

Class	Equiv_1	Equiv_2	Equiv_3	Equiv_4	Reduct
Equiv_1	Null	c	a	$a \vee b$	Reduct_1
Equiv_2	c	Null	$a \vee c$	$a \vee b \vee c$	Reduct_2
Equiv_3	a	$a \vee c$	Null	b	Reduct_3
Equiv_4	$a \vee b$	$a \vee b \vee c$	b	Null	Reduct_4

TABLE 2.6   Simplified Decision Table

Class	Price (a)	Horsepower (b)	User Age (c)	Recommendation (d)
Equiv_1	High	×	Old	Car type 1
Equiv_2	×	×	Young	Car type 2
Equiv_3	Medium	Medium	×	Car type 2
Equiv_4	×	High	×	Car type 3

Step 6. Use the reduct to simplify the decision table. Now the reduct attributes set could be applied to make reductions on the equivalence classes. For example, since Reduct_1 is $a \wedge c$ (or $\{a,c\}$), the price and user age are sufficient to differentiate Equiv_1 from all other classes. Therefore, the horsepower attribute is removed from the table. The simplified version of the table is shown in Table 2.6. After this table is obtained, only a subset of the original attributes is needed to differentiate the types of car. Decision efficiency is thus improved.

2.3.3   Prototypical Case Generation Using Reducts with Fuzzy Representation

In this section we illustrate how rough set theory and the concept of equivalence classes (i.e., objects that are indiscernible using the available attributes) can be applied to select the prototypical cases for building a CBR system. This set of prototypical cases will be indexed and retrieved at later stages of the CBR reasoning tasks.

Salamo and Golobardes [7], Pal and Mitra [8], and Polkowskim et al. [9] have developed different reduction techniques to identify the prototypical patterns. Therefore, given a case base, if one can identify that only one element of the equivalence class is needed to represent the entire class, enormous storage space will be saved. The other possible criterion to consider for reduction is to keep only those case features that preserve the indiscernibility relation. Since the attributes rejected are redundant, their removal will not affect the task of case retrieval based on feature similarities. There are usually several subsets of case features that can preserve indiscernibility, and the minimal subsets among them are called *reducts*. Finding a minimal reduct is a NP-*hard problem* [10], which means that finding this is a nontrivial task requiring considerable computation. With this in mind, let us describe here an approach developed by Pal and Mitra [8,11] for selecting prototypical cases based on rough-fuzzy integration.

Selection and generation of cases can be regarded as the two important phases in building a good case base for a CBR system. Whereas case selection deals with selecting informative prototypes from the data, case generation concerns the construction of cases that need not necessarily include all the data points given. Rough set theory was developed for classificatory analysis of data tables. The main goal of rough set theoretical analysis is to synthesize an approximation (upper and lower) of the classifying concepts from the target data. Whereas fuzzy set theory assigns to each object a grade of belongingness to represent an imprecise set, rough set theory focuses on the ambiguity caused by limited discernibility of objects within a given domain. The key concepts here are those of information granule and reducts. An *information granule* formalizes the concept of finite-precision representation of objects in real-life situations, and *reducts* represent the *core* of an information system (in terms of both objects and features) in a granular universe. It may be noted that cases also represent the informative and irreducible part of a problem. Hence, rough set theory is a natural choice for case selection in domains that are data rich, which contain uncertainties and allow tolerance for imprecision. Additionally, rough sets have the capability to handle complex objects (e.g., proofs, hierarchies, frames, rule bases), thereby extending the applicability of rough CBR systems. Rough mereology, allowing for tolerance relations, has been used to design approximate CBR systems [9]. Here rough sets assist in fast case retrieval by circumventing the adaptation step. Recently, rough and fuzzy sets have been integrated within a soft computing framework, the aim being to develop a (synergistic) model of uncertainty that is stronger than using either technique alone [12].

In the rough-fuzzy case generation method of Pal and Mitra [8,11], each pattern (object) is represented by its fuzzy membership with respect to overlapping linguistic property sets: low, medium, and high (i.e., an object originally having n features is now represented by $3n$ variables). This type of representation is quite versatile in representing a wide class of data: for example, linguistic, set form, and numeric [12,13]. In terms of their attributes, discernibility of the fuzzified objects is then computed in the form of a discernibility matrix. Using rough set theory, a number of decision rules are generated from the discernibility matrix. The rules represent *rough clusters* of points in the original feature space. The fuzzy membership functions, corresponding to the region modeled by a rule, are then stored as a case. A strength factor representing the a priori probability (size) of the cluster is associated with each case. To summarize, each case contains fuzzy linguistic sets (parameters of the membership functions) corresponding to the attributes appearing in the reducts, and the strength factor. In the *retrieval phase*, these fuzzy membership functions are used to compute the similarity of stored cases with an unknown pattern.

It may be noted that unlike most case selection schemes, the cases generated by this algorithm need not include any of the actual objects encountered; rather, they represent regions that are either in the original feature space or else in a reduced space. More important, cases are represented by subsets of features (attributes) that are of different sizes. This type of variable and reduced length representation of cases will decrease the retrieval time. Furthermore, since this algorithm deals only with the information granules, not the actual data set, its significance in data mining applications is evident.

The effectiveness of this methodology is demonstrated on some real-life data sets, including those that have large dimension and size. Cases are evaluated in terms of the classification accuracy obtained using the 1-NN rule. Comparisons are made with the conventional IB3 algorithm [14] and random case selection method. This methodology is found to perform better in terms of 1-NN accuracy, as well as improving case generation time and average case retrieval time.

In the following sections we describe the methodology for fuzzy linguistic representation of objects. Then we present the methodology used to obtain dependency rules and the algorithm for mapping dependency rules to cases. The case retrieval mechanism is then described. Finally, some experimental results and comparisons with other approaches are presented.

2.3.3.1 Linguistic (Fuzzy) Representation of Patterns

As mentioned earlier, rough set theory deals with a set of objects in a granular universe. Here we describe a way of obtaining the granular feature space using fuzzy linguistic representation of patterns. Only the case of numeric features is mentioned here. (Features in descriptive and set forms can also be handled in this framework.) Details of the methodologies involved may be found in [12,13].

Let a pattern (object) $\hat{e}$ be represented by n numeric features (attributes) (i.e., $\hat{e} = [F_1, F_2, \ldots, F_n]$). Each feature is described in terms of its fuzzy membership values, corresponding to three linguistic fuzzy sets: low (L), medium (M), and high (H). Thus, an n-dimensional pattern vector is represented as a $3n$-dimensional vector [12,13]:

$$\hat{e} = [\mu_{\text{low}(F_1)}(\hat{e}), \mu_{\text{medium}(F_1)}(\hat{e}), \mu_{\text{high}(F_1)}(\hat{e}), \mu_{\text{low}(F_2)}(\hat{e}), \mu_{\text{medium}(F_2)}(\hat{e}), \mu_{\text{high}(F_2)}(\hat{e}), \ldots,$$

$$\mu_{\text{low}(F_n)}(\hat{e}), \mu_{\text{medium}(F_n)}(\hat{e}), \mu_{\text{high}(F_n)}(\hat{e})] \tag{2.11}$$

where $\mu_{\text{low}(F_j)}(\cdot)$, $\mu_{\text{medium}(F_j)}(\cdot)$, and $\mu_{\text{high}(F_j)}(\cdot)$ indicate the membership values of $(\cdot)$ to the fuzzy sets low, medium, and high along feature axis j, $\mu(\cdot) \in [0,1]$.

For each input feature F_j, the fuzzy sets "low," "medium," and "high" are characterized individually by a π-membership function whose form is [13]

$$\mu(F_j) = \pi(F_j, c, \lambda) = \begin{cases} 2\left(1 - \dfrac{||F_j - c||}{\lambda}\right) & \text{for} \quad \frac{\lambda}{2} \leq ||F_j - c|| \leq \lambda \\ 1 - 2\left(\dfrac{||F_j - c||}{\lambda}\right)^2 & \text{for} \quad 0 \leq ||F_j - c|| \leq \frac{\lambda}{2} \\ 0 & \text{otherwise} \end{cases} \tag{2.12}$$

where λ (>0) is the radius of the π-function with c as the central point. For each of the fuzzy sets *low*, *medium*, and *high*, λ and c take different values. These values are chosen so that the membership functions for these three fuzzy sets have overlapping nature (intersecting at membership value 0.5; see Fig. 2.10).

Let us now explain the procedure for selecting the centers (c) and radii (λ) of the overlapping π-functions. Let m_j be the mean of the pattern points along the jth axis. Then m_{jl} and m_{jh} are defined as the mean (along the jth axis) of the pattern points having coordinate values in the range $(F_{j\ min}, m_j)$ and $(m_j, F_{j\ max})$, respectively, where

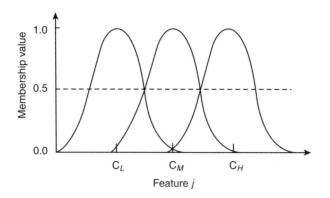

Figure 2.10 Membership functions for linguistic property sets low (L), medium (M), and high (H) for each feature axis.

$F_{j\,max}$ and $F_{j\,min}$ denote the upper and lower bounds of the dynamic range of feature F_j. The centers and the radii of the three π-functions are defined as

$$
\begin{aligned}
c_{\text{low}(F_j)} &= m_{jl}\\
c_{\text{medium}(F_j)} &= m_j\\
c_{\text{high}(F_j)} &= m_{jh}\\
\lambda_{\text{low}(F_j)} &= c_{\text{medium}(F_j)} - c_{\text{low}(F_j)}\\
\lambda_{\text{high}(F_j)} &= c_{\text{high}(F_j)} - c_{\text{medium}(F_j)}\\
\lambda_{\text{medium}(F_j)} &= 0.5\left(c_{\text{high}(F_j)} - c_{\text{low}(F_j)}\right)
\end{aligned}
\tag{2.13}
$$

Here we take into account the distribution of the pattern points along each feature axis while choosing the corresponding centers and radius of the linguistic fuzzy sets.

The aforesaid three overlapping functions along each axis generate the fuzzy granulated feature space in n dimensions. The granulated space contains 3^n granules with fuzzy boundaries among them. Here the granules (clumps of similar objects or patterns) are attributed by the three fuzzy linguistic values "low," "medium," and "high." The degree of belongingness of a pattern to a granule (or the degree of possessing a property low, medium, or high by a pattern) is determined by the corresponding membership function.

Furthermore, if one wishes to obtain crisp granules (or crisp subsets), α-cut, $0 < \alpha < 1$, of these fuzzy sets may be used. (The α-cut of a fuzzy set is a crisp set of points for which membership values are greater than or equal to α.)

2.3.3.2 Dependency Rule Generation A principal task in the method of rule generation is to compute reducts relative to a particular kind of information system (the decision system). Relativized versions of these matrices and functions are the basic tools used in this computation. The d-reducts and d-discernibility matrices are used for this purpose [10]. The methodology is described below.

Let $Ts = \langle U, A \rangle$ be a decision table, and let Tc and $Td = \{d_1^*, d_2^*, \ldots, d_l^*\}$ be the sets of condition and decision attributes, respectively. Divide the decision table $Ts = \langle U, A \rangle$ into l tables $Ts_i = \langle U_i, A_i \rangle, i = 1, 2, \ldots, l$, corresponding to the l decision attributes $d_1^*, d_2^*, \ldots, d_l^*$, where $U = U_l \cup \cdots \cup U_l$ and $A_i = Tc \cup \{Td_i\}$.

Let $\{x_{i1}, x_{i2}, \ldots, x_{ip}\}$ be the set of those objects of U_i that occur in $Ts_i, i = 1, 2, \ldots, l$. Now for each d_i-reduct, let $B = \{b_1, b_2, \ldots, b_k\}$. A discernibility matrix [denoted $\boldsymbol{M}_{d_i}(B)$] from the d_i-discernibility matrix is defined as follows:

$$c_{ij} = \{\text{Attr} \in B : \text{Attr}(x_i) \neq \text{Attr}(x_j)\} \qquad \text{for } i, j = 1, 2, \ldots, n \qquad (2.14)$$

For each object $x_j \in \{x_{i1}, x_{i2}, \ldots, x_{ip}\}$, the discernibility function $f_{d_i^*}^{x_j}$ is defined as

$$f_{d_i^*}^{x_j} = \wedge\{\vee(c_{ij}) : 1 \leq i, j \leq n, j < n, c_{ij} \neq \emptyset\} \qquad (2.15)$$

where $\vee(c_{ij})$ is the disjunction of all members of c_{ij}. Then $f_{d_i^*}^{x_j}$ is converted to its conjunctive normal form. Dependency rules r_i are obtained [i.e., $d_i^* \leftarrow g_i$, where g_i is the disjunctive normal form of $f_{d_i^*}^{x_j}, j \in \{i_1, i_2, \ldots, i_p\}$]. The dependency factor df_i for r_i is given by

$$df_i = \frac{\text{card}(\text{POS}_i(d_i^*))}{\text{card}(U_i)} \qquad (2.16)$$

where $\text{POS}_i(d_i^*) = \cup_{X \in I_{d_i^*}} l_i(X), l_i(X)$ is the lower approximation of X with respect to $I_{d_i^*}$. For real-life data sets, the decision tables are often inconsistent, and perfect decision rules may not be obtained. In these situations the degree of precision of an imperfect rule is quantified by the dependency factor.

2.3.3.3 Case Generation

Here we describe a methodology [8,11] for case generation on the fuzzy granulated space as obtained in previous sections. This involves two tasks: generation of fuzzy rules using rough set theory, and mapping the rules to cases. Since rough set theory operates on crisp granules (i.e., subsets of the universe), we need to convert the fuzzy membership values of the patterns to binary values or to convert the fuzzy membership functions to binary functions in order to represent crisp granules (subsets) for application of rough set theory. This conversion can be done using an α-cut.

The schematic diagram for the generation of cases is shown in Figure 2.11. One may note that the inputs to the case generation process are fuzzy membership functions, the output cases are also fuzzy membership functions, but the intermediate

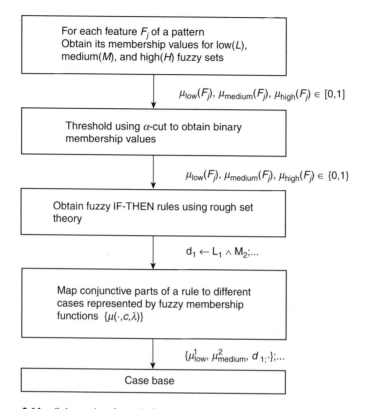

For each feature F_j of a pattern
Obtain its membership values for low(L),
medium(M), and high(H) fuzzy sets

$\mu_{low}(F_j)$, $\mu_{medium}(F_j)$, $\mu_{high}(F_j) \in [0,1]$

Threshold using α-cut to obtain binary
membership values

$\mu_{low}(F_j)$, $\mu_{medium}(F_j)$, $\mu_{high}(F_j) \in \{0,1\}$

Obtain fuzzy IF-THEN rules using rough set
theory

$d_1 \leftarrow L_1 \wedge M_2;...$

Map conjunctive parts of a rule to different
cases represented by fuzzy membership
functions $\{\mu(\cdot,c,\lambda)\}$

$\{\mu_{low}^1, \mu_{medium}^2, d_{1,\cdot}\};...$

Case base

Figure 2.11   Schematic of rough-fuzzy case generation and retrieval. (From [11].)

rough set–theoretic processing is performed on binary functions representing crisp sets (granules). For example, the inputs to block 2 are fuzzy membership functions. Its outputs are binary membership functions that are used for rough processing in blocks 3 and 4. Finally, the outputs of block 4, representing cases, are again fuzzy membership functions. Each task is discussed below.

Consider the $3n$ fuzzy membership values of an n-dimensional pattern $\hat{e}$. Then select only those attributes having values greater than or equal to Th ($= 0.5$, say). In other words, we obtain a 0.5-cut of all the fuzzy sets to obtain binary membership values corresponding to the sets "low," "medium," and "high." After the binary membership values are obtained for all the patterns, we constitute a decision table for rough set rule generation. As the method considers multiple objects in a class, a separate $n_k \times 3n$-dimensional attribute–value decision table is generated for each class d_k^* (where n_k indicates the number of objects in d_k^*). Let there be m sets O_1, $O_2, \ldots, O_m$ of objects in the table that have identical attribute values, and card(O_i) $= n_{ki}$, $i = 1, 2, \ldots, m$, such that $n_{k1} \geq n_{k2} \geq \cdots \geq n_{km}$ and $\sum_{i=1}^{m} n_{ki} = n_k$. The attribute–value table can now be represented as an $m \times 3n$ array.

Let $n_{k1'}, n_{k2'}, \ldots, n_{km'}$ denote the distinct elements among $n_{k1}, \ldots, n_{km}$ such that $n_{k1'} > n_{k2'} > \cdots > n_{km'}$. Let a heuristic frequency threshold function be defined as [15]

$$\mathrm{Tr} = \left\lceil \frac{\sum_{i=1}^{m} 1 / \left(n_{k_i'} - n_{k_{i+1}'}\right)}{\mathrm{Th}} \right\rceil \tag{2.17}$$

such that all entries that have a frequency of less than Tr are eliminated from the table, resulting in a reduced attribute–value table. Note that the main motive for introducing this threshold function lies in reducing the size of the case base. Attempts can be made to eliminate noisy pattern representatives (having lower values of n_{ki}) from the reduced attribute–value table. Rough dependency rules are generated from this reduced attribute–value table using the methodology described in Section 2.3.3.2.

2.3.3.4 *Mapping Dependency Rules to Cases*

We now describe a technique for mapping rough dependency rules to cases. This algorithm is based on the observation that each dependency rule (having a frequency above some threshold) represents a cluster in the feature space. It may be noted that only a subset of features appear in each of the rules, which indicates that the entire feature set is not always necessary to characterize a cluster. A *case* is constructed from a *dependency rule* in the following manner:

1. Consider the antecedent part of the rule: Split it into atomic formulas containing only a conjunction of literals.
2. For each atomic formula, generate a case that contains the centers and radius of the fuzzy linguistic variables (low, medium, and high) that are present in the formula. (Thus, multiple cases may be generated from a rule.)
3. Associate with each case, generated the precedent part of the rule and the case strength equal to the dependency factor of the rule [equation (2.16)]. The strength factor reflects the size of the corresponding cluster and the significance of the case.

Thus a case has the following structure:

```
case{
Feature i: fuzzset_i: center, radius;
......
Class k
Strength
}
```

where *fuzzset* denotes the fuzzy linguistic variables "low," "medium," and "high." The method is illustrated below with the help of an example. One may note that while 0.5-cut is used to convert the $3n$ fuzzy membership functions of a pattern to binary functions for rough set rule generation, the original fuzzy functions are

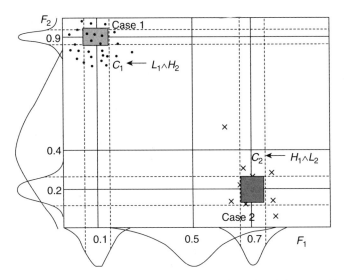

Figure 2.12 Rough-fuzzy case generation for a two-dimensional data set. (From [11].)

retained in order to use them to represent the cases generated. These are also illustrated in Figure 2.11, where the outputs μ^1_{low} and μ^2_{medium} are fuzzy sets.

As an example, consider a data set having two features F_1 and F_2 and two classes. Let us assume that the following two dependency rules are obtained from the reduced attribute table:

$$\text{class}_1 \leftarrow L_1 \wedge H_2, \text{df} = 0.5,$$
$$\text{class}_2 \leftarrow H_1 \wedge L_2, \text{df} = 0.4,$$

where L_1, L_2, H_1, and H_2 mean that F_1 is low, F_2 is low, F_1 is high, and F_2 is high, respectively; and df represents the dependency factor.

Let the parameters of the fuzzy linguistic sets "low," "medium," and "high" be as follows:

Feature 1 : $c_L = 0.1, \lambda_L = 0.5, c_M = 0.5, \lambda_M = 0.7, c_H = 0.7, \lambda_H = 0.4.$

Feature 2 : $c_L = 0.2, \lambda_L = 0.5, c_M = 0.4, \lambda_M = 0.7, c_H = 0.9, \lambda_H = 0.5.$

This example is illustrated in Figure 2.12. We have the following two cases:

```
case 1 {
Feature No: 1, fuzzset (L): center = 0.1, radius = 0.5
Feature No: 2, fuzzset (H): center = 0.9, radius = 0.5
Class = 1
Strength = 0.5
}
```

```
case 2 {
Feature No: 1, fuzzset (H): center = 0.7, radius = 0.4
Feature No: 2, fuzzset (L): center = 0.2, radius = 0.5
Class = 2
Strength = 0.4
}
```

2.3.3.5 Case Retrieval

Each case obtained in Section 2.3.3.4 is a collection of fuzzy sets {*fuzzsets*} described by a set of one-dimensional π-membership functions with different c and λ values. To compute the similarity of an unknown pattern $\hat{e}$ (of dimension n) to a case e_p (of variable dimension n_p, $n_p \le n$), we use

$$\text{SM}(\hat{e}, e_p) = \sqrt{\frac{1}{n_p} \sum_{j=1}^{n_p} [\mu_{fuzzset}^j(F_j)]^2} \qquad (2.18)$$

where $\mu_{fuzzset}^j(F_j)$ is the degree of belongingness of the jth component of $\hat{e}$ to *fuzzset* representing the case e_p. When $\mu^j = 1$ for all j, SM $(\hat{e}, e_p) = 1$ (maximum) and when $\mu^j = 0$ for all j, SM$(\hat{e}, e_p) = 0$ (minimum). Therefore, equation (2.18) provides a collective measure computed over the degree of similarity of each component of the unknown pattern with the corresponding one of a stored case. The higher the value of the similarity, the closer pattern $\hat{e}$ is to the case e_p. Note that fuzzy membership functions in equation (2.18) take care of the distribution of points within a granule, thereby providing a better similarity measure between $\hat{e}$ and e_p than the conventional Euclidean distance between two points.

For classifying (or to provide a label to) an unknown pattern, the case closest to the pattern in terms of SM$(\hat{e}, e_p)$ is retrieved and its class label is assigned to the pattern. Ties are resolved using the parameter *Case Strength*.

2.3.3.6 Results and Comparison

Experiments were performed [8,11] on the following three real-life data sets available in the UCI Machine Learning Archive [16]:

- *Forest cover type:* contains 10 dimensions, seven classes, and 586,012 samples. It is a geographical information system (GIS) data set that represents forest cover type (pine, fir, etc.) in the United States. The variables used are cartographic and remote sensing measurements. The variables are numeric.
- *Multiple features:* consists of features of handwritten numerals (0 through 9) that were extracted from a collection of Dutch utility maps. There are 2000 patterns, 649 features (all numeric), and 10 classes.
- *Iris:* contains 150 instances, four features, and three classes for iris. The features are all numeric.

Class 1 ← $L_1 \wedge H_2 \wedge L_3$	df = 0.81
Class 2 ← $M_1 \wedge L_2 \wedge M_4$	df = 0.81
Class 3 ← $H_1 \wedge H_4$	df = 0.77

Figure 2.13   Rough dependency rules for the iris data.

The cases generated using the rough-fuzzy methodology described in Section 2.3.3.5 were compared with those obtained using the following three case selection methodologies:

- Instance-based learning algorithm, IB3 [14]
- Instance-based learning algorithm, IB4 [17]. Here the feature weights are learned using random hill climbing. This method selects a specified number of features that have high weights
- Random case selection

```
case 1 {
Feature No: 1, fuzzset (L): center = 5.19, radius = 0.65
Feature No: 2, fuzzset (H): center = 3.43, radius = 0.37
Feature No: 3, fuzzset (L): center = 0.37, radius = 0.82
Class = 1
Strength = 0.81
}

case 2 {
Feature No: 1, fuzzset (M): center = 3.05, radius = 0.34
Feature No: 2, fuzzset (L): center = 1.70, radius = 2.05
Feature No: 4, fuzzset (M): center = 1.20, radius = 0.68
Class = 2
Strength = 0.81
}

case 3 {
Feature No: 1, fuzzset (H): center = 6.58, radius = 0.74
Feature No: 4, fuzzset (H): center = 1.74, radius = 0.54
Class = 3
Strength = 0.77
}
```

Figure 2.14   Cases generated for the iris data.

TABLE 2.7   Comparison of Case Selection Algorithms for Iris Data

Algorithm	Number of Cases	n_{avg}	Classification Accuracy	e_{gen} (sec)	t_{ret} (sec)
Rough-fuzzy	3	2.67	98.17	0.2	0.005
IB3	3	4	98.00	2.50	0.01
IB4	3	4	90.01	4.01	0.01
Random	3	4	87.19	0.01	0.01

Comparison was performed on the basis of the following quantities:

- 1-NN classification accuracy using the cases where 10% of the entire samples were used as the training set for case generation, while the remaining 90% were used as the test set
- Number of cases stored in the case base
- Total CPU time required for case generation
- Average CPU time required to retrieve a case for the patterns in the test set

For illustration, we first present for the iris data, the rough dependency rules, and the corresponding cases (representing three granular regions) generated from these rule, as shown in Figures 2.13 and 2.14. Here classes 1, 2, and 3 are *Iris setosa*, *Iris versicolor*, and *Iris virginica*, respectively. Four features are sepal length (F_1), sepal width (F_2), petal length (F_3), and petal width (F_4). Comparative results for the different case generation methodologies are presented in Tables 2.7, 2.8, and 2.9 for the *iris*, *forest cover type*, and *multiple features* data sets, respectively. These results compare the number of cases, the 1-NN classification accuracy, the average number of features per case (n_{avg}) and the case generation (e_{gen}) and retrieval (t_{ret}) times. The results in the tables show that the cases obtained using the rough-fuzzy method [8,11] are much superior to both the random selection method and the IB4 method, and that these cases are close to IB3 in terms of classification accuracy. It should be noted that the rough-fuzzy method requires significantly less time for case generation than do IB3 and IB4. The tables also show that the average number of features stored per case (n_{avg}) using the rough-fuzzy technique is much lower than the original data dimension (n); as a direct consequence, its average retrieval time is very low. The IB4 method also stores cases with a reduced number of features

TABLE 2.8   Comparison of Case Selection Algorithms for Forest Data

Algorithm	Number of Cases	n_{avg}	Classification Accuracy	e_{gen} (sec)	t_{ret} (sec)
Rough-fuzzy	542	4.10	67.01	244	4.4
IB3	545	10	66.88	4055	52.0
IB4	545	4	50.05	7021	4.5
Random	545	10	41.02	17	52.0

TABLE 2.9   Comparison of Case Selection Algorithms for Multiple Features Data

Algorithm	Number of Cases	n_{avg}	Classification Accuracy	e_{gen} (sec)	t_{ret} (sec)
Rough-fuzzy	50	20.87	77.01	1096	10.05
IB3	52	649	78.99	4112	507
IB4	52	21	41.00	8009	20.02
Random	50	649	50.02	8.01	507

and has a low retrieval time; however it has lower accuracy than that of the rough-fuzzy method. Moreover, unlike the rough-fuzzy method, all the stored cases have an equal number of features.

Since rough set theory is used to obtain cases through crude rules, case generation time is reduced. Also, since only the informative regions and relevant features are stored, case retrieval time decreases significantly. The algorithm is suitable for mining data sets, which are large both in dimension and size, where the requirement is to find approximate but effective solutions quickly.

Note that there are some algorithms [18] that consider a reduced feature representation of the cases but the dimension of the cases is fixed, unlike the rough-fuzzy method described here, which generates cases with variable dimensions. Although a particular rough set method is used for the task of providing the dependency rules [8,11], any other version of rough set methods can be used [19].

2.4   CASE INDEXING

2.4.1   Traditional Indexing Method

In the traditional relational database approach, indexes refer to the primary or secondary keys of a record. *Indexing* refers to the task of mapping the record key to the storage location. It could be done by using direct access methods, such as hashing, or indexed methods, such as building a B^+-tree (see Fig. 2.15) or an R-tree (i.e., range tree) for organizing the records on hand. Searching and retrieving of records to determine their locations are done either by mapping them to the index tree or by using a hashing algorithm. For example, the way of mapping the records to a B^+-tree can be explained by referring to Figure 2.15. Here the nodes at the bottom

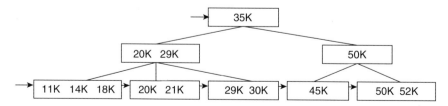

Figure 2.15   Example of B^+-tree for indexing household incomes.

layer of the tree are leaf nodes and those in the two layers above are inner nodes. Inner nodes contain the value or a value interval of the index, and leaf nodes contain the pointers to the storage locations of cases. An inner node may generate three child nodes. The upper boundary of the interval in its left child node is smaller than the lower boundary of its parent, and the lower boundary of the right child is equivalent or larger than the upper boundary of its parent. The lower boundary of the middle child is equivalent or larger than that of its parent, and its upper boundary is smaller than that of its parent. In this case, a feature such as family income is taken as the index. For example, given a new case with a family income of 11K, we first compare this index value with the value of 35K in the root node of the tree. Since 11K is smaller than 35K, we go next to the left branch of the root node. This time it is found that 11K is smaller than the lower boundary of the interval in the node, so we visit the left branch of this node where the storage location of the given case could be retrieved. Similarly, if the index value of the given case is 21K, smaller than 35K and falling into the interval 20K to 29K, the location of the given case could be found in the middle child node of the node [20K, 29K].

Thus, organizing cases using this B^+-tree structure requires a precise matching of the primary key attribute. For example, assume that the searching target has an index value of 12K; although the record with index 11K is just next to it, it will not be retrieved. Other improved index structures, such as R-trees, R*-trees, and R^+-trees, support range and multidimensional searching of multimedia records; however, they are still based on the concept of exact matching or crisp boundary. That is, the objects are either within the range or outside the range; overlapping of concepts is not allowed. Unlike the traditional way of retrieving records from database, the retrieval methods incorporated with soft computing techniques follow a method of similarity-based case retrieval that is content aware. The searching result might not be the record that is obtained by matching the target exactly, but will be the one(s) most similar to the target. The definition of the index is also generalized to a new level. Here the index refers to some general concept or object rather than a crisp one, so that it can accommodate many records with varying degrees of possibility as possible solutions. For example, consider a layered neural network as an index where the mapping of the query case (input) to a group of possible solutions (output) is made based on approximate matching, unlike traditional indexing, which involves mapping based only on exact matching. Moreover, since the neural network is using a generalized concept, it needs to be learned through training before use. Like ANN, a Bayesian model and prototype-based incremental neural network can be used as an index under a soft computing framework for indexing different cases. It should be mentioned here that the Bayes model takes care of the uncertainty arising from the randomness of events and provides decisions with varying probabilities. These are described below.

2.4.2 Case Indexing Using a Bayesian Model

Suppose that the sample space S has n classes $C_1, C_2, \ldots, C_n$ of events, where $C_i \cap C_j = \emptyset, i \neq j, i, j = 1, 2, \ldots, n$, and $\sum_{i=1}^{n} P(C_i) = 1$, where $P(C_i)$ is the

TABLE 2.10   Fourteen Cases of the Car-Buying Example

Case_ID	User Age	Price	Horsepower	Fuel Capacity	Recommendation
1	Young	High	Powerful	Medium	Car type 1
2	Young	High	Powerful	Large	Car type 1
3	Old	Low	Common	Large	Car type 1
4	Young	Medium	Common	Medium	Car type 1
5	Old	Medium	Powerful	Large	Car type 1
6	Middle Age	High	Powerful	Medium	Car type 2
7	Old	Medium	Powerful	Medium	Car type 2
8	Old	Low	Common	Medium	Car type 2
9	Middle Age	Low	Powerful	Large	Car type 2
10	Young	Low	Common	Medium	Car type 2
11	Old	Medium	Common	Medium	Car type 2
12	Young	Medium	Common	Large	Car type 2
13	Middle Age	Medium	Powerful	Large	Car type 2
14	Middle Age	High	Common	Medium	Car type 2

probability of occurrence of C_i. Given an event X defined on S, if the class conditional probabilities $P(X|C_i)$, $i = 1, 2, \ldots, n$, are known, then for each C_i, the a posteriori probability of $P(C_i|X)$, $i = 1, 2, \ldots, n$, is defined by the Bayes formula as

$$P(C_i|X) = \frac{P(X|C_i)P(C_i)}{P(X)} = \frac{P(X|C_i)P(C_i)}{\sum_{i=1}^{n} P(X|C_i)P(C_i)} \qquad (2.19)$$

Consider our earlier car-buying example consisting of 14 cases, as shown in Table 2.10. Each case has four features: user age, price, horsepower, and fuel capacity, and a class label, car type 1 or car type 2. Given a query case X: user age = "young," price = "medium," horsepower = "common," fuel capacity = "medium," the objective is to identify its class label. The Bayes formula can thus be used as follows. Let C_1 and C_2 denote the cases of car type 1 and car type 2, respectively. $C_1 \cap C_2 = \emptyset$, $C_1 \cup C_2 = S$, where S is the sample space (or case base). Then the solution is $X \in C_j$ if

$$P(C_j|X) = \max_i(P(C_i|X)), \qquad i, j = 1, 2 \qquad (2.20)$$

Since the denominator of equation (2.19) is constant, we can write that the solution is $X \in C_j$ if

$$P(X|C_j)P(C_j) = \max_i(P(X|C_i)P(C_i)), \qquad i = 1, 2 \qquad (2.21)$$

To compute $P(X|C_i)$, the attributes are assumed to be conditionally independent. Therefore,

$$P(X|C_i) = \prod_{j=1}^{4} P(x_j|C_i) \qquad (2.22)$$

where x_j is the jth ($j = 1, 2, 3, 4$) feature of X, and $X = \{x_1, x_2, x_3, x_4\} = \{$user age, price, horsepower, fuel capacity$\}$.

Using the case base in Table 2.10, the class label of X can be identified as a solution using the following algorithm:

Step 1. Compute $P(C_1)$ and $P(C_2)$.

$$P(C_1) = P(\text{recommendation} = \text{``car type 1''}) = \frac{|C_1|}{|S|} = \frac{5}{14} = 0.36$$

$$P(C_2) = P(\text{recommendation} = \text{``car type 2''}) = \frac{|C_2|}{|S|} = \frac{9}{14} = 0.64$$

Step 2. Decompose the query case X, and compute $P(x_i|C_1)$ and $P(x_i|C_2)$, $i = 1, 2, 3, 4$.

$P(x_1|C_1) = P(\text{user age} = \text{``young''}|\text{recommendation} = \text{``car type 1''}) = \frac{3}{5} = 0.6$

$P(x_2|C_1) = P(\text{price} = \text{``medium''}|\text{recommendation} = \text{``car type 1''}) = \frac{2}{5} = 0.4$

$P(x_3|C_1) = P(\text{horsepower} = \text{``common''}|\text{recommendation} = \text{``car type 1''}) = \frac{2}{5} = 0.4$

$P(x_4|C_1) = P(\text{fuel capacity} = \text{``medium''}|\text{recommendation} = \text{``car type 1''}) = \frac{2}{5} = 0.4$

$P(x_1|C_2) = P(\text{user age} = \text{``young''}|\text{recommendation} = \text{``car type 2''}) = \frac{2}{9} = 0.22$

$P(x_2|C_2) = P(\text{price} = \text{``medium''}|\text{recommendation} = \text{``car type 2''}) = \frac{4}{9} = 0.44$

$P(x_3|C_2) = P(\text{horsepower} = \text{``common''}|\text{recommendation} = \text{``car type 2''}) = \frac{5}{9} = 0.56$

$P(x_4|C_2) = P(\text{fuel capacity} = \text{``medium''}|\text{recommendation} = \text{``car type 2''}) = \frac{6}{9} = 0.67$

Step 3. Compute $P(X|C_i) = \prod_{j=1}^{4} P(x_j|C_i)$, $i = 1, 2$.

$$P(X|C_1) = \prod_{j=1}^{4} P(x_j|C_1) = P(X|\text{recommendation} = \text{``car type 1''})$$

$$= 0.6 \times 0.4 \times 0.4 \times 0.4 = 0.0384$$

$$P(X|C_2) = \prod_{j=1}^{4} P(x_j|C_2) = P(X|\text{recommendation} = \text{``car type 2''})$$

$$= 0.22 \times 0.44 \times 0.56 \times 0.67 = 0.036$$

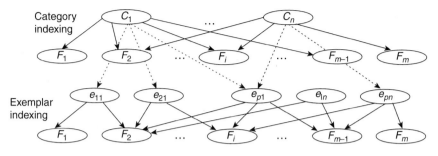

Figure 2.16 Two-layered Bayesian network for indexing category and exemplar.

Step 4. Compute and choose the largest value of $P(X|C_i)P(C_i)$, $i = 1, 2$.

$$P(X|C_1)P(C_1) = P(X|\text{recommendation}$$
$$= \text{``car type 1''})P(\text{recommendation} = \text{``car type 1''})$$
$$= 0.0384 \times 0.36 = 0.014$$
$$P(X|C_2)P(C_2) = P(X|\text{recommendation}$$
$$= \text{``car type 2''})P(\text{recommendation} = \text{``car type 2''})$$
$$= 0.036 \times 0.64 = 0.023$$
$$P(X|C_2)P(C_2) > P(X|C_1)P(C_1)$$

The maximum value of $P(C_i|X)$ is 0.023 with category C_2. So the input case falls into C_2 and hence the solution is: Recommend car type 2.

The Bayes formula has been used for indexing cases by Rodríguez et al. [20] (see Fig. 2.16). The case base is organized into N categories and each category has several representative cases (i.e., exemplars). Each of these exemplars is represented by a set of features and depicts a set of "similar" cases. For example, category C_1 contains the cases with feature F_1 to feature F_{m-1} and an exemplar of C_1 (e.g., e_{11} is represented with two features, F_1 and F_2).

Given a new case *nc*, the first layer of the network is used as a category index to rank [using equation (2.19)] the location of the probable categories (in descending order) to which the new case may belong (see Fig. 2.17). The second layer is then used as an exemplar index to determine from the aforesaid categories in descending order the location of a group of cases (represented by an exemplar e_i) that are most similar to the new case. To measure such similarity between e_i and *nc*, equation (2.19) is used and then it is checked if the corresponding $P(e_i|nc)$ exceeds a predefined threshold. The algorithm stops as soon as an exemplar satisfying the threshold is found. The category index thus narrows down the searching space for the exemplar index, thereby improving the efficiency of the indexing structure. (Note that it may sometimes happen that the best-match exemplar is found in a less probable

```
Classify(nc, e_c, CL)
```

Input: A new case described by a set of features $nc = \{F_a, \ldots, F_i\}$
Results: The category list CL, and the examplar e_c that is "simliar"
The following local variables are used:
H is a list sorted in categories
E is a list sorted in exemplars
C_c is the current category
done is a Boolean variable

Step 1. Rank the Categories
Compute conditional probabilities in the first BN
    (i.e., the category features BN)
Compute $P(C_j|nc)$ $\forall$ $C_j \in$ CL
Set H to the list of categories ranked in descending order of $P(C_j|nc)$

Step 2. Determination of an Exemplar
$e_c = nil$
$C_c = first(H)$; the list H returns $\emptyset$ at the end
Done = false
while (not *done*) and $(C_c \neq \emptyset)$ do begin
Compute conditional probabilities in the second BN
    (i.e., the exemplar features BN)
Compute $P(e_i|nc)$ $\forall e_i \in C_c$
Set E to the list of exemplars in C_c ranked in descending
    order of $P(e_i|nc)$
$e_c = first(E)$
if $P(e_c|nc) >$ threshold, then
done = true
else
$C_c = next(H)$
end(if)
end(while)
if *done*, then
CL = all categories that contain (e_c)
else
$e_c = nil$
CL $= \emptyset$
end(if)

Figure 2.17   Classification algorithm. (From [20].)

category rather than a higher one, since the threshold criterion is not satisfied for the latter.)

2.4.3 Case Indexing Using a Prototype-Based Neural Network

There are three principal advantages of using neural networks to index cases: (1) as neural networks are inherently a parallel architecture, the index developed could be used to improve the case searching and retrieval efficiency; (2) once a neural network is trained successfully, it is very robust in handling noisy input case data with incomplete or missing information; and (3) neural networks are particularly suitable in situations where learning the class labels of cases is performed incrementally by accumulating data. Some of the basic models of artificial neural networks and characteristic features are described in Appendix B.

Similar to the concept of an exemplar described in the Bayesian network in Section 2.4.2, the idea of prototype-based indexing is to construct a set of prototypes that could be used to index a group of similar cases. The construction of these prototypes can be achieved through learning from the case data using a neural network. One of these examples is the prototype-based incremental neural network developed by Malek [21,22] (see Fig. 2.18). The network here has three layers. The input layer contains one unit for each feature x_i, $i = 1, 2, \ldots, n$. The output layer contains one unit for each class C_i, $i = 1, 2, \ldots, m$. Each class contains a group of similar cases. A unit in the hidden layer is considered as a prototype and is represented by a vector [e.g., the third prototype is represented as $w_3 = (w_{13}, w_{23}, \ldots, w_{n3})$, where w_{i3} is the average value of the ith ($i = 1, 2, \ldots, n$) component of all the cases that are represented by this prototype.

During the case retrieval phase, given a new current case, the task of case retrieval is performed by finding out which prototype in the hidden layer should be

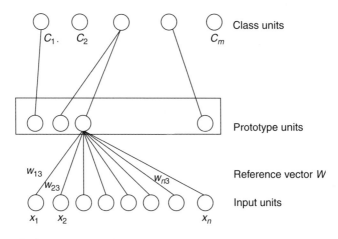

Figure 2.18 General architecture of a prototype-based network. (From [21].)

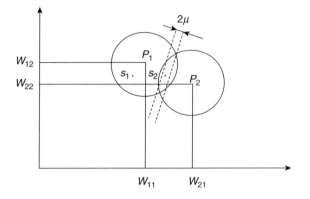

Figure 2.19 Two prototypes activated by S_2. (From [21].)

activated. For instance, if the first prototype w_1 is activated among the group of prototypes, then class C_1 to which w_1 refers should be located. To determine which prototype should be activated, a similarity measure $SM(X,w_i)$ that is derived from the Euclidean distance and a user-specified similarity threshold is used. First, the similarities between the present case and the prototypes are calculated, and then among the prototypes whose similarity values are above the threshold, the one with the maximum similarity is activated.

However, the retrieval task mentioned above cannot be performed effectively without a sufficient number of prototypes in the hidden layer. To build this layer for indexing cases, an incremental learning algorithm is used. Three types of modification of the prototype layer are possible:

1. *Construction of a new prototype.* Two types of situations can lead to the construction of a new prototype. First, if there is initially no prototype for comparison in the hidden layer, the training case is then added to this layer as a prototype. An influence region of this prototype could be defined with a radius that is initialized to be equal to the user-specified threshold. After the prototype is created, a new region of memory is allocated and linked to it. Second, if a sufficient number of cases are found in the *atypical memory*, [i.e., memory consisting of those cases that could activate more than one prototype (see Figs. 2.19 and 2.20), or activate no prototype], they are clustered, and for each cluster a new prototype is generated.

2. *Specialization and expansion of the prototypes.* If an input training case is misclassified by an activated prototype, the influence region of this prototype should be reduced to exclude the training case from its class. By doing so the prototype is specialized. Meanwhile, the influence region of another prototype that should have included the training case will be expanded so that it could be activated.

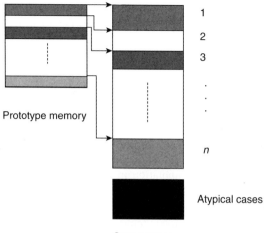

Figure 2.20   Two levels of case indexing structure. (From [21].)

3. *Tuning of prototypes.* If the training case is classified correctly, it is added to the linked memory of the guiding prototype. By averaging the feature values of all similar cases in the same group, a new prototype that is more representative is generated and used to replace the old one.

The process of learning will not stop until the atypical region stabilizes, which means that any new input training case can activate one and only one prototype in the hidden layer. P_1 and P_2 in Figure 2.19 are two prototypes that are associated with different classes. S_2 is a case that falls into an uncertain region and the network is unable to give a decision because two different prototypes are activated, so this case is added to the atypical memory.

2.4.4   Case Indexing Using a Three-Layered Back-Propagation Neural Network

An approach for maintaining feature weights in a dynamic context was developed by Yang et al. [23] using a layered network. This network can be used as an index to locate a set of potential solutions that are learned (extracted) from the previous cases. Given a new case, its suggested solution can be obtained by adapting the said potential solutions, thus reducing the need to compare all the cases in the case base. In this model the first layer represents the feature values of the cases, the second layer represents the problem contexts, and the third layer represents the solutions that are extracted from the cases. The feature value layer is connected to the problem layer through a set of weights denoted by V. Similarly, the problem layer is connected to the solution layer through another set of weights, denoted by W.

The weight training algorithm of the three-layered back-propagation neural network model has the following steps. Given the input feature values, a problem score is computed as

$$\text{Sp}_j = \frac{2}{1 + \exp[-\lambda \sum_{i=1}^{I} (V_{ji} X_i)]} - 1 \tag{2.23}$$

where Sp_j is the score of the problem Pr_j and $j = 1, 2, 3, \ldots, J$. V_{ji} is a weight attached to the problem Pr_j and the feature input FV_i. The problem score outputs are then forwarded to the solution layer as inputs. Computation of the score of a solution is again similar to computation in the problem layer. After the score of a solution is computed, it will be shown to the user for his or her judgment. If the user is not satisfied with the solution score, they can specify the desired score. This information will be used in the computation of errors for weight learning using a back-propagation process. The computation of the delta (Δ) values is first done at the solution layer to update the weights in that layer. They are computed as

$$\Delta S_k = \tfrac{1}{2}(D_{S_k} - S_k)(1 - S_k^2) \tag{2.24}$$

where ΔS_k is the learned delta value for solution S_k and D_{S_k} is the desired score for S_k. These learned delta values are then propagated back to the problem layer. The learned delta value for the problem Pr_j can be computed as

$$\Delta \text{Pr}_j = \frac{1}{2}(1 - \text{Sp}_j^2) \sum_{k=1}^{K} (\Delta S_k W_{kj}) \tag{2.25}$$

where the weight W_{kj} is attached to the connection between a solution S_k and a problem Pr_j. These weights are modified or adjusted using the delta values [computed with equation (2.23)] as

$$W_{k,j}^{\text{new}} = W_{k,j}^{\text{old}} + \eta \Delta S_k \text{Sp}_j \tag{2.26}$$

where $W_{k,j}^{\text{new}}$ is the new weight to be computed, $W_{k,j}^{\text{old}}$ the old weight attached to the connection between S_k and problem Pr_j, and η the learning rate. The weights attached between the feature value layer and the problem layer is computed in a similar manner.

2.5 SUMMARY

In this chapter we present both the traditional and soft computing approaches to case representation and case indexing. For case representation, we first explain the relational model, the object-oriented model, and the predicate logic model. A methodology for generation of prototypical cases from a fuzzy granulated space is then described for case representation. This uses rough sets in generating rules through reducts. These prototypical cases can also be used as indexes for retrieving the relevant cases at later stages of the CBR reasoning tasks. Traditional indexing models such as B^+-tree and R-tree can provide an effective way of indexing cases. We show with examples how a Bayesian probability model can be used for indexing. Other soft computing methods described include those based on a prototype-based incremental neural network and a three-layered back-propagation neural network. It may be mentioned here that fuzzy layered networks (e.g., in [13]) and knowledge-based networks [24] can be used for efficient indexing, particularly when the input is described in terms of imprecise information such as fuzzy hedges, set values, and missing components. Although the present literature on using the GA for indexing is scarce, its potential in determining the best index structure through searching is evident.

REFERENCES

1. J. Kolodner, *Case-Based Reasoning*, Morgan Kaufmann, San Francisco, 1993.

2. P. Koton, Reasoning about evidence in causal explanation, in *Proceedings of the Seventh National Conference on Artificial Intelligence (AAAI-88)*, St. Paul, MN, vol. 1, pp. 256–261, 1988.

3. W. Dubitzky, A. Schuster, J. Hughes, and D. Bell, An advanced case-knowledge architecture based on fuzzy objects, *Applied Intelligence*, vol. 7, pp. 187–204, 1997.

4. Z. Pawlak, *Rough Sets: Theoretical Aspects of Reasoning about Data*, Kluwer Academic, Norwell, MA, 1991.

5. W. P. Ziarko (ed.), *Rough Sets, Fuzzy Sets and Knowledge Discovery*, Springer-Verlag, Berlin, 1994.

6. T. Y. Lin and N. Cercone (eds.), *Rough Sets and Data Mining: Analysis of Imprecise Data*, Kluwer Academic, Norwell, MA, 1996.

7. M. Salamo and E. Golobardes, Rough sets reduction techniques for case-based reasoning, in *Proceedings of the Fourth International Conference on Case-Based Reasoning (ICCBR-01)*, Vancouver, British Columbia, Canada, Springer-Verlag, Berlin, pp. 467–482, 2001.

8. S. K. Pal and P. Mitra, Case generation: a rough-fuzzy approach, in *Proceedings of the Fourth International Conference on Case-Based Reasoning (ICCBR-01)*, Vancouver, British Columbia, Canada, Springer-Verlag, Berlin, pp. 236–242, 2001.

9. L. Polkowskim, A. Skowron, and J. Komorowski, Approximate case-based reasoning: a rough mereological approach, in *Proceedings of the Fourth German Workshop on Case-Based Reasoning, System Deveolpment and Evaluation*, vol. 55 (H. D. Burkhard and M. Lenz, eds.), Humboldt University, Berlin, pp. 144–151, 1996.

10. A. Skowron and C. Rauszer, The discernibility matrices and functions in information systems, in *Intelligent Decision Support: Handbook of Applications and Advances of the Rough Sets Theory* (R. Slowinski, ed.), Kluwer Academic, Dordrecht, The Netherlands, pp. 331–362, 1992.

11. S. K. Pal and P. Mitra, Case generation using rough sets with fuzzy representation, *IEEE Transactions on Knowledge and Data Engineering*, vol. 16, no. 3, 2004.

12. S. K. Pal and A. Skowron (eds.), *Rough Fuzzy Hybridization: New Trends in Decision Making*, Springer-Verlag, Singapore, 1999.

13. S. K. Pal and S. Mitra, Multi-layer perceptron, fuzzy sets and classification, *IEEE Transactions on Neural Networks*, vol. 3, pp. 683–697, 1992.

14. D. W. Aha, D. Kibler, and M. Albert, Instance-based learning algorithms, *Machine Learning*, vol. 6, pp. 37–66, 1991.

15. M. Banerjee, S. Mitra, and S. K. Pal, Rough fuzzy MLP: knowledge encoding and classification, *IEEE Transactions on Neural Networks*, vol. 9, pp. 1203–1216, 1998.

16. C. Blake and C. Merz, UCI Repository of machine learning databases, *http://www.ics. uci.edu/$\sim$mlearn/MLRepository.html*, University of California–Irvine, Department of Information and Computer Sciences, 1998.

17. D. W. Aha, Tolerating noisy, irrelevant, and novel attributes in instance-based learning algorithms, *International Journal of Man–Machine Studies*, vol. 36, pp. 267–287, 1992.

18. D. R. Wilson and T. R. Martinez, Reduction techniques for instance-based learning algorithms, *Machine Learning*, vol. 38, no. 3, pp. 257–286, 2000.

19. M. Szczuka, Rough sets and artificial neural networks, in *Rough Sets in Knowledge Discovery 2: Applications, Case Studies and Software Systems* (L. Polkowski and A. Skowron, eds.), Physica-Verlag, Heidelberg, pp. 449–470, 2000.

20. A. F. Rodríguez, S. Vadera, and L. E. Sucar, A probabilistic model for CBR, in *Proceedings of the Second International Conference on Case-Based Reasoning (ICCBR-97)*, Providence, RI, Springer-Verlag, Berlin, pp. 623–632, 1997.

21. M. Malek, A connectionist indexing approach for CBR systems, in *Proceedings of the First International Conference on Case-Based Reasoning (ICCBR-95)*, Sesimbra, Portugal, Springer-Verlag, Berlin, pp. 520–527, 1995.

22. M. Malek, Hybrid approaches for integrating neural networks and case-based reasoning: from loosely coupled to tightly coupled models, in *Soft Computing in Case-Based Reasoning* (S. K. Pal, T. S. Dillon, and D. S. Yeung, eds.), Springer-Verlag, London, pp. 73–94, 2000.

23. Q. Yang, J. Wu, and Z. Zhang, Maintaining large case bases using index learning and clustering, in *Proceedings of the International Young Computer Scientist Conference 1999*, Nanjing, China, vol. 1, pp. 466–473, 1999.

24. S. Mitra, R. K. De, and S. K. Pal, Knowledge-based fuzzy MLP for classification and rule generation, *IEEE Transactions on Neural Networks*, vol. 8, pp. 1338–1350, 1997.

CHAPTER 3

CASE SELECTION AND RETRIEVAL

3.1 INTRODUCTION

Case selection and retrieval is usually regarded as the most important step within the case-based reasoning cycle. This problem has been studied in both supervised and unsupervised frameworks by many researchers and practitioners, and hundreds of different algorithms and various approaches have been developed to retrieve similar cases from the case base. In this process the similarity measures adopted in a CBR system will greatly influence retrieval performance. One of the most important assumptions in case-based reasoning is that similar experiences can guide future reasoning, problem solving, and learning. This similarity assumption [1] is used in problem-solving and reasoning systems when target problems are dealt with by resorting to a previous situation with common conceptual features. Most of today's case matching and retrieval algorithms are based on this assumption, and this leads to the development of various case clustering and classification techniques. Among these techniques, weighted feature-based similarity is the most common form of method used in computing relatedness among cases. If there are too many cases in a case base, a set of representative cases may be selected beforehand (thereby signifying the importance of case selection), so that the tasks of matching and retrieval will be managed efficiently. However, there are an increasing number of arguments about using this simple feature-based similarity as an explanation of human thinking and categorization [1–5]. This is because similarity cannot be studied in an isolated manner, and there are many possible assumptions and constraints

Foundations of Soft Case-Based Reasoning. By Sankar K. Pal and Simon C. K. Shiu
ISBN 0-471-08635-5 Copyright © 2004 John Wiley & Sons, Inc.

that could affect measurements of similarity among cases. For example, similarity may be interpreted as relatedness among problem features (i.e., the problem space) or relatedness among solution features (i.e., the solution space). Questioning on the similarity assumption opens up the possibility of using many other alternatives for defining similarity, such as adaptation-guided retrieval [1], similarity with confidence level [6], context-guided retrieval [7], and similarity among structured case representations [8]. In this chapter we deal with techniques based on the similarity assumption, especially with how soft computing tools can be used to extend traditional feature-based similarity computation. However, readers should bear in mind that similarity may mean different things to different people, or different things to the same person at different times.

The rest of this chapter is organized as follows. In Section 3.2 we review the concept of similarity measure. The use of fuzzy sets for similarity assessment and its computation is explained in Section 3.3. Fuzzy classification and clustering methods for case retrieval are described in Section 3.4. The significance of case feature weighting (importance) and the determination of weights using derivative- and GA-based optimization methods are covered in Section 3.5. In Section 3.6 we address the tasks of case selection and retrieval using neural networks. In Section 3.7 a supervised neuro-fuzzy method for case selection that is especially suitable for selecting cases from overlapping regions has been explained along with experimental results. Finally, in Section 3.8 we describe a method of integrating rough sets, fuzzy sets, and self-organizing networks for extracting cases in an unsupervised mode of learning for developing a compact case base. Its merits in terms of learning time, cluster quality, and compact representation of data are demonstrated.

3.2 SIMILARITY CONCEPT

The meaning of similarity always depends on the underlying context of a particular application, and it does not convey a fixed characteristic that applies to any comparative context. In CBR, the computation of similarity becomes a very important issue in the case retrieval process. The effectiveness of a similarity measurement is determined by the usefulness of a retrieved case in solving a new problem. Therefore, establishing an appropriate similarity function is an attempt at handling the deeper or hidden relationships between the relevant objects associated with the cases. There are (broadly) two major retrieval approaches [9]. The first is based on the computation of distance between cases, where the most similar case is determined by evaluation of a similarity measure (i.e., a metric). The second approach is related more to the representational/indexing structures of the cases. The indexing structure can be traversed to search for a similar case. We describe next the basic concepts and features of some of the distance measures used in this regard.

3.2.1 Weighted Euclidean Distance

The most common type of distance measure is based on the location of objects in Euclidean space (i.e., an ordered set of real numbers), where the distance is calculated as the square root of the sum of the squares of the arithmetical differences between the corresponding coordinates of two objects. More formally, the weighted Euclidean distance between cases can be expressed in the following manner. Let $CB = \{e_1, e_2, \ldots, e_N\}$ denote a case library having N cases. Each case in this library can be identified by an index of the corresponding features. In addition, each case has an associated action. More formally, we use a collection of features $\{F_j \ (j = 1, 2, \ldots, n)\}$ to index the cases and a variable V to denote the action. The ith case e_i in the library can be represented as an $(n + 1)$-dimensional vector, that is, $e_i = (x_{i1}, x_{i2}, \ldots, x_{in}, \theta_i)$, where x_{ij} corresponds to the value of feature F_j $(1 \leq j \leq n)$ and θ_i corresponds to the value of action V $(i = 1, 2, \ldots, N)$.

Suppose that for each feature F_j $(1 \leq j \leq n)$, a weight w_j $(w_j \in [0, 1])$ has been assigned to the jth feature to indicate the importance of that feature. Then, for any pair of cases e_p and e_q in the library, a weighted distance metric can be defined as

$$d_{pq}^{(w)} = d^{(w)}(e_p, e_q) = \left[\sum_{j=1}^{n} w_j^2 (x_{pj} - x_{qj})^2\right]^{1/2} = \left(\sum_{j=1}^{n} w_j^2 \chi_j^2\right)^{1/2} \tag{3.1}$$

where $\chi_j^2 = (x_{pj} - x_{qj})^2$. When all the weights are equal to 1, the weighted distance metric defined above degenerates to the Euclidean measure $d_{pq}^{(1)}$, in short, is denoted by d_{pq}. Using the weighted distance, a similarity measure between two cases, $SM_{pq}^{(w)}$, can be defined as

$$SM_{pq}^{(w)} = \frac{1}{1 + \alpha d_{pq}^{(w)}} \tag{3.2}$$

where α is a positive constant. The higher the value of $d_{pq}^{(w)}$, the lower the similarity between e_p and e_q. When all of the weights take a value of 1, the similarity measure is denoted by $SM_{pq}^{(1)}$, $SM_{pq}^{(1)} \in [0, 1]$.

It should be noted that the real-valued features discussed above could be extended without difficulty to the features that take values in a normed vector space. For example, assume that for each feature, a distance measure has already been defined. The distance measure for the jth feature is denoted by ρ_j; that is, ρ_j is a mapping from $F_j \times F_j$ to $[0, \infty]$ (where F_j denotes the domain of the jth feature) with the following properties:

(a) $\rho_j(a, b) = 0$ if and only if $a = b$.
(b) $\rho_j(a, b) = \rho_j(b, a)$.
(c) $\rho_j(a, b) \leq \rho_j(a, c) + \rho_j(c, b)$.

For numerical and nonnumerical features, some typical formulas for the distance measure, such as the following, could be used:

(a) $\rho_j(a, b) = |a - b|$ if a and b are real numbers.

(b) $\rho_j(A, B) = \max_{a \in A, b \in B} |a - b|$ if A and B are intervals.

(c) $\rho_j(a, b) = \begin{cases} 1 & \text{if } a \neq b \\ 0 & \text{if } a = b \end{cases}$ if a and b are symbols.

In these circumstances, the distance between two cases e_p and e_q can be computed by

$$d_{pq}^w = \sqrt{\sum_{j=1}^{n} w_j^2 \rho_j^2 (e_{p_j}, e_{q_j})} \qquad (3.3)$$

3.2.2 Hamming and Levenshtein Distances

Hamming distance was originally conceived for detection and correction of errors in digital communication [10]. It is simply defined as the number of bits that are different between two bit vectors. For example, $d(0, 1) = 1, d(001, 011) = 1,$ $d(000, 111) = 3, d(111, 111) = 0$. In the field of artificial intelligence, there is often a need to search through large state spaces. Therefore, a predefined state space metric would be very useful to guide the search, and this metric is usually referred to as the *Hamming distance*. In the context of case-based reasoning, an evaluation function (or an algorithm) is used to compute the Hamming distance between a query case and the cases in the case library. The case that has the minimum Hamming distance is selected as the most similar case to the query case. In many applications, the Hamming distance is represented by a state-space graph, and the computation of the distance involves algorithms that traverse the arcs and nodes of this graph.

Levenshtein distance is a measure of the similarity between two strings and is defined as the number of deletions, insertions, or substitutions required to transform the source string into the target string. The Levenshtein distance is also called the *edit distance*. Each transformation is associated with a cost, and the objective is to find the cheapest way to transform one string into another. There are many possible applications of Levenshtein distance, for example, it could be used in measuring errors in text entry tasks [11] as well as for comparing sequences.

3.2.3 Cosine Coefficient for Text-Based Cases

In many practical applications, comparing text-based cases is necessary. This leads to a requirement for similarity metrics that can measure the relation among documents. In the field of information retrieval (IR), cluster analysis has been used to create groups of documents that have a high degree of association between members from the same group and a low degree between members from different groups. To cluster the documents in a data set, a distance measure is used. A variety of distance and similarity measures in this regard are mentioned [12]. Among them, the Dice, Jaccard, and Cosine coefficients have the dual attractions of simplicity and normalization.

Let DT $= \{s_1, s_2, \ldots, s_N\}$ denote an N-tuple of documents s_i, where s_i is a member of the set DT. Let TT $= \{t_1, t_2, \ldots, t_M\}$ denote an M-tuple of term types (i.e., specific words or phrases) t_j, where t_j is a member of the set TT. The term frequency, denoted by TF_{s_i, t_j} is the frequency of occurrence of term t_j in document s_i. The inverse document frequency, denoted by $\text{IDF}_{\text{DT}, t_j}$, provides high values for rare words and low values for common words. It is defined as the logarithm (to base 2) of the ratio of the number of the documents in set DT to the number of documents s_i in set DT that contains at least one occurrence of term t_j. Each term is then assigned a score, called the *weight* of the term, which is defined as $W_{s_i, t_j} = \text{TF}_{s_i, t_j} \times \text{IDF}_{\text{DT}, t_j}$.

This score has a higher value if a term is both frequent in relevant documents and infrequent in the document collection as a whole. However, longer documents may be given a greater weight because of a higher term frequency. Therefore, a process of normalization is necessary. The *normalized weight* is defined as

$$\text{NW}_{s_i, t_j} = \frac{W_{s_i, t_j}}{\sqrt{\sum_{j=1}^{M} \left(\text{TF}_{s_i, t_j}\right)^2 \left(\text{IDF}_{\text{DT}, t_j}\right)^2}} \tag{3.4}$$

Finally, the similarity of two documents S_i and S_j can be expressed as a cosine relationship (cosine coefficient):

$$\text{Cos SM}_{S_i, S_j} = \frac{\sum_{j=1}^{M} \left(W_{s_i, t_j} W_{s_j, t_j}\right)}{\sqrt{\sum_{j=1}^{M} \left(W_{s_i, t_j}\right)^2 \sum_{j=1}^{M} \left(W_{s_j, t_j}\right)^2}} \tag{3.5}$$

3.2.4 Other Similarity Measures

Here we provide some more useful similarity measures. A survey in this regard is available in [9]. Let SM_{pq} denote a similarity measure between two cases, a new query case e_p and a stored case e_q. A similarity measure that is based on the ratio model has been proposed by Tversky [13]:

$$\text{SM}_{pq} = \frac{\alpha(\text{common})}{\alpha(\text{common}) + \beta(\text{different})} \tag{3.6}$$

where "common" and "different" represent the number of attributes that are similar or dissimilar, respectively, between the new query case e_p and the stored case e_q. Usually, this decision involves referring to a threshold value so that features are classified as similar if their similarity is above that threshold. The value of α and β are the corresponding weights whose values could be determined by an expert or by using machine-learning techniques.

A similarity measure that is based on the number of production rules that are instantiated has been proposed by Sebay and Schoenauer [14]:

$$SM_{pq} = \sum_i w(r_i) \tag{3.7}$$

where (r_i) represents the rules that are learned from the case base and w is the weight assigned. A similarity measure based on the contrast model [13] is proposed by Weber [15]:

$$SM_{pq} = \alpha f(e_p \cap e_q) - \beta f(e_p - e_q) - \gamma f(e_q - e_p) \tag{3.8}$$

where e_p represents a new query case and e_q represents a stored case. The intersection $(e_p \cap e_q)$ describes those attributes that are common to e_p and e_q, and the complement sets $(e_p - e_q)$ and $(e_q - e_p)$ describe those attributes that are observed only in the query case (not in the stored case) and only in the stored case (not in the query case), respectively. f denotes some operator or algorithm to compute the matching score of the related sets. α, β, and γ are the corresponding weights.

Several other similarity metrics are also proposed. These take into consideration different comparative features, such as the number of consecutive matches [16], the degree of the normalized associations between attributes [17], the "typicality" of cases [18], the relevance of certain attributes between a new query case and a stored case [19], the degree of similarity in the relationships between attributes [20], structural similarities [21], similarity based on object-oriented class hierarchy [22], and supervised and unsupervised fuzzy similarity measures [23].

3.2.5   *k*-Nearest Neighbor Principle

The k-nearest neighbor principle involves search for the k nearest cases to the current input case using a distance measure [such as Euclidean distance; see equation (3.1)] and then selecting the class of the majority of these k cases as the retrieval one. In other words, for classification of the query case, the confidence for each class is computed as m_i/k, where m_i is the number of cases among the k nearest cases that belong to class i. The class with the highest confidence is then assigned to the query case. Usually, to improve the chances of a correct decision for the present cases that are near the boundary between two classes, a threshold β is set so that at least β among the k nearest neighbors have to agree on the classification. Although the k-NN algorithm is simple, it does suffer from a major disadvantage: when the number of feature dimensions and the number of cases in the case base are large, the computation required for classification is enormous [24,25].

3.3   CONCEPT OF FUZZY SETS IN MEASURING SIMILARITY

Fuzzy set theory was introduced by Lotfi Zadeh [26] in 1965. Since then it has been an active area of research for many scientists and engineers, and there have been tremendous advances and new developments both theoretically and in applications [27,28]. Basic concepts and definitions of fuzzy set theory are given in Appendix A for the convenience of readers.

In Section 3.2, the concept and forms of traditional similarity were introduced. It is easy to discover that traditional similarity, such as weighted Euclidean distance, can only handle features with real-value and characteristic feature values. However, in the real situation, case features are often incompletely or uncertainly specified. For example, one of the features of cases in a CBR system may be described by such linguistic terms as *low, medium,* and *high*. Then for implementing the process of case matching and retrieval, one needs to define an appropriate metric of similarity. The traditional definition of similarity is obviously not valid and at least not effective to deal with this difficulty. Here the concept of fuzzy set provides a good tool to handle the problem in a natural way. In fuzzy set theory, we may consider the linguistic term as a fuzzy number, which is a type of fuzzy set. Then a membership function is determined with respect to the given linguistic term. When a real value of the feature of a given problem is input, the corresponding values of membership to different linguistic terms are obtained through the membership functions. That is, after approximate matching, the real-valued features are transformed to linguistic features. Then, depending on the problem, to select the best-matching case or the best set of cases, one needs to define some similarity measures and algorithms for computing fuzzy similarity. Before we define them, we provide a mathematical framework that signifies the relevance of fuzzy similarity in case matching.

3.3.1 Relevance of Fuzzy Similarity in Case Matching

Case-based reasoning is based on the idea that if two problems are similar with respect to a set of attributes S_1 that describe them, they could still be similar with respect to another set of attributes, S_2. Then we can take advantage of past experience with problems that are similar (with respect to S_1) to a current problem (concerning S_2) to infer a plausible value for S_2. Davies and Russell [29] used functional dependencies between sets of attributes S_1 and S_2 to control the inference process. However, this framework is very restrictive since inference is possible only when both of the following conditions hold: (1) there is a perfect identity between the S_1 values describing the two problems, and (2) when the S_1 values are equal, the S_2 values are also equal (i.e., the functional dependency "S_1 determines S_2" holds). Later, Gilboa and Schmeidler [30] advocated a similarity-based approach where a case is described as a triple (situation, act, result) and a decision maker's nonnegative utility function assigns a numerical value θ to result Rt. When faced with a new situation St_0, the decision maker is supposed to choose an act Ac that maximizes the following equation (CU *maximization*), a counterpart of classical expected utility theory used in decision making under uncertainty:

$$CU_{St_0}(Ac) = \sum_{(St,Ac,Rt)\,\in\,M} SM(St_0, St)\theta(Rt) \qquad (3.9)$$

where SM is a nonnegative function that estimates the similarity between the situations. SM denotes the similarity between the current situation, St_0, and those already encountered and stored in the memory M. Moreover, it is assumed that $\forall$ St and $\forall$ Ac, $\exists$ Rt such that $(St, Ac, Rt) \in M$, which means that results are uniquely determined by the act applied in the context of a given problem. It is also assumed that $\forall$ St, $\exists$ Ac such that $(St, Ac, Rt) \in M$ and $\theta(Rt) \neq 0$, which means that only the *best* act in the context of s is stored in the memory. Gilboa and Schmeidler also gave an axiomatic derivation of this CU *maximization*, within a formal model.

Later, Dubois et al. [31] provided a fuzzy set theoretic approach to case-based decisions. In this approach, "strong" dependencies express that the value of S_1 (fuzzily) determines the value of S_2, and "weak" dependencies state only that if the S_1 *values* are similar, it is possible that the S_2 *values* are similar. The details of this approach are as follows. Let M denote a memory of cases encountered which are represented by pairs (St_i, θ_i) for $i = 1, 2, \ldots, n$, where St_i denotes a problem and θ_i represents the associated outcome (solution). The current problem will be denoted by St_0 and its intended solution by θ_0.

A fuzzy functional dependency of the form "the more similar St_1 is to St_2, the more similar θ_1 is to θ_2," where (St_1, θ_1) and (St_2, θ_2) are cases in M, can be modeled by the constraint

$$(St_1, \theta_1), (St_2, \theta_2) \in M, FS(St_1, St_2) \leq FT(\theta_1, \theta_2) \qquad (3.10)$$

where FS and FT are fuzzy proximity relations. FS and FT are symmetric and reflexive. In fact, the constraint (3.10) expresses that when St_1 and St_2 are close, then θ_1 and θ_2 should be at least as close as them. Moreover, if FT is defined such that $FT(\theta_1, \theta_2) = 1 \Leftrightarrow \theta_1 = \theta_2$, the classical functional dependency $St_1 = St_2 \Rightarrow \theta_1 = \theta_2$ is a consequence of constraint (3.10) using the reflexivity of FS. In this case, constraint (3.10) is then clearly stronger than a classical functional dependency.

Let $(St, \theta) \in M$, and (St_0, θ_0) be the current situation; then we have

$$FS(St, St_0) \leq FT(\theta, \theta_0)$$

where θ_0 is unknown. Thus, the constraint defines a set of possible values for θ_0, namely, $\{\theta_0 \in FT \mid FS(St, St_0) \leq FT(\theta, \theta_0)\}$. Since it applies for any (St, θ) in M, we obtain the following set B of possible values for θ_0:

$$B = \cap(St, \theta)_{\in M}\{\theta' \in FT \mid FS(St, St_0) \leq FT(\theta, \theta')\} \qquad (3.11)$$

Note that B may be empty if FT is not permissive enough. It can be shown that the nonemptiness of B can be guaranteed by the coherence in the set of fuzzy gradual rules so that the more St_0 is FS-*similar* to St_i, the more θ_0 should be FT-*similar* to θ_i for each case $(St_i, \theta_i) \in M$.

The requirement that the value of St uniquely determines the value of θ, or the requirement that at least when St_i and St_j are close, θ_i and θ_j should also be close, may be felt too strong in some practical applications. Memory M may, for instance, simultaneously include cases such as (St, θ) and (St, θ') with θ different from θ'. In such a situation, we can use a weaker version of the principle underlying case-based reasoning and state this as, "the more similar St_1 is to St_2, the greater is the possibility that θ_1 and θ_2 are similar." The formal expression of this principle requires clarification of the intended meaning of *possibility*. Rules of the form "the more S_1 is A, the greater the possibility that S_2 is B" correspond to a particular kind of fuzzy rule called *possibility rules* [31]. They express that "the more S_1 is A, the greater the possibility that B is a range for S_2," which can be understood as "$\forall u, if\ S_1 = u,$ it is possible at least at the degree $A(u)$ that S_2 *lies in* B." When B is an ordinary subset, it clearly expresses that (1) if $v \in B$, v is possible for S_2 at least at the level $A(u)$ *if* $S_1 = u$, and (2) if $v \notin B$, nothing can be said about the minimum possibility level of the value v for S_2. This definition leads to the following constraint on the conditional possibility distribution $P_{S_2|S_1}$ representing the rule

$$\forall u \in U, \forall\ v \in V, \min(A(u), B(v)) \le P_{S_2|S_1}(v, u) \qquad (3.12)$$

for a full justification of its semantics when both A and B are fuzzy.

Applying the principle "the more similar St and St_0 are (in the sense of S), the more possible it is that θ and θ_0 are similar (in the sense of FT)," the fuzzy set of possible values θ' for θ_0 is given by

$$P_{\theta_0}(\theta') \ge \min(\text{FS}(St, St_0), \text{FT}(\theta, \theta')) \qquad (3.13)$$

As can be seen, what we obtain is the fuzzy set of values θ' that are FT-similar to θ, "truncated" by the global degree $\text{FS}(St, St_0)$ of the similarity between St and St_0. Since it applies to all pairs $(St, \theta) \in M$, we obtain the following fuzzy set B of possible values θ' for θ_0 $[B(\theta') = P_{\theta_0}(\theta')]$:

$$B(\theta') = \max_{(St, \theta) \in M} \min(\text{FS}(St, St_0), \text{FT}(\theta, \theta')) \qquad (3.14)$$

Inference based on a fuzzy case rule [29] can be divided into two stages. In the first stage, an inference is based on how well the facts of a new case correspond to the elements associated with a (precedent) case rule. This is judged using a criterion yes or no, which is evaluated according to the degree of fuzzy membership between the facts and elements. In the second stage, the inference from the precedent case to the new case is drawn, and this is directed by the similarity between the cases. The conclusions obtained from both these stages are compared with that of the precedent case. If they are identical with the conclusion of the precedent case, the new case has the same result as the precedent. If they are not identical with that conclusion, a decision concerning the new case cannot be supported by the precedent.

When we make a judgment on the correspondence between the facts of the new case and the elements of a (precedent) case rule (that is represented by the fuzzy membership function), a yes or no judgment is unnecessary for inference by case rule. Accordingly, the center of gravity of the fuzzy membership function of these cases can be defined as

$$CG(A_i) = \frac{\int_{c_1}^{c_2} x\mu_{A_i}(x)\, dx}{\int_{c_1}^{c_2} \mu_{A_i}(x)\, dx} \tag{3.15}$$

where $U = [c_1, c_2]$, A_i is the fuzzy set that describes the judgment on the correspondence between the elements of a case rule (i) and the facts of a new case. μ_{A_i} is the membership function of A_i. $CG(A_i)$ lies in $[0, 1]$. Considering 0.5 as the threshold, if the value of the center of gravity is greater (or less) than 0.5, the judgment is yes (or no).

The *distance* between two centers of gravity, $|CG(A) - CG(B)|$, is used to describe the degree of similarity. To satisfy the conditions of similarity relations, the degree of similarity $SM(A, B)$ is calculated using

$$SM(A, B) = 1 - |CG(A) - CG(B)| \tag{3.16}$$

The conceptual similarity of an elemental item within the cases is assessed as

$$\Delta SM = e^{-\beta \Delta d^2} \tag{3.17}$$

where $\beta (\beta > 0)$ denotes amendment accuracy, which should be fixed beforehand. The formulation of the provision acceptance depends on the elemental item that belongs to this issue j. The value Δd is the distance between the relevant items from the two cases (e_p, e_q), and it can be computed as

$$\Delta d = |CG(e_p) - CG(e_q)| \tag{3.18}$$

The similarity of the issue j is assessed using the similarity of the associated elemental items as

$$SM_j = \min\{\Delta SM_1, \Delta SM_2, \ldots, \Delta SM_i, \ldots, \Delta SM_n\}, \quad \Delta SM_i \in [0, 1], \quad n \in N \tag{3.19}$$

where n is the number of elemental items that belong to the issue j.

As a general rule, more than one issue can be compared between two cases. The algorithm applied when there is more than one relevant issue should also be considered. In this situation, a weight w_i is introduced into the case-based retrieval. The average similarity is then weighted. It is calculated as

$$\overline{SM} = \frac{\sum (w_i SM_i)}{\sum w_i} \tag{3.20}$$

Let each frame of a precedent case and a new case be represented as follows:

$$\text{Precedent}: \quad A = \{A_i\}_{i=1}^n$$
$$\text{New case}: \quad B = \{B_i\}_{i=1}^n$$

where A is the frame that represents the precedent, B the frame that represents the new case, A_i the fuzzy set that describes the judgment concerning the elements of the precedent case rule, and n the quantity of slots in a frame.

The similarity assessment is performed as follows: Let the membership functions of A_i and B_i be μ_{A_i} and μ_{B_i}, respectively. The center of gravity of A_i and B_i can be computed using equation (3.15). Let $SM(A_i, B_i)$ be the degree of similarity between A_i and B_i. Then the degree of similarity between A and B can be obtained from

$$SM(A, B) = \min(SM(A_1, B_1), \ldots, SM(A_n, B_n)) \tag{3.21}$$

If the degree of similarity is greater than the threshold (which was determined in advance), the conclusion is that frame B is the same as frame A. For example, if there is a conclusion that "the proposal is sufficiently definite" in a precedent, the conclusion of new case is also "the proposal is sufficiently definite." If the degree of similarity is less than the given threshold, the conclusion is that frame B cannot arrive at the same conclusion as that of A. This does not necessarily mean that the new case has an opposite conclusion to the precedent. Perhaps it is possible to reach the same conclusion using another precedent.

This system has been used for drawing inferences from contract law [32], where the judgment associated with the elements of the case rule was represented by the concepts of membership and vagueness in fuzzy theory. The inference based on the case rule was made by the yes or no judgment and the degree of similarity. The yes or no judgment was made according to the center of gravity of the membership function. The degree of similarity was calculated using fuzzy matching. It was on the basis of these requirements that the inference experiment was carried out.

3.3.2 Computing Fuzzy Similarity between Cases

There are several methods [33] for computing the similarity between cases:

- Numeric combination of feature vectors (properties, attributes), representing the known cases, using different combination rules.
- Similarity of structured representations, in which each case is represented as a structure, such as a directed graph, and thus the similarity measure takes into account the structure of the different attributes of the case and not only the attribute value.
- Goal-driven similarity assessment, in which the attributes of the cases that are to be compared with those of a new case depend on the goal sought. This

means that some attributes of a case are not important in the light of a certain goal and thus should not be taken into account in the similarity calculation.

- Rule-based similarity assessment, in which the cases in the case base (CB) are used to create a set of rules on the feature vector of the cases. This rule set is then used to compare the cases in the CB and to solve the new case (e.g., if exactly the same rules fire for exactly the same attributes in both an old case and a new one, these cases are said to be very similar).

- Aggregation of the foregoing methods according to application-specific hierarchies.

The detailed steps in constructing the fuzzy case and the similarity computation can be summarized [34] in the following way:

Step 1: Case representation. In this representation, a vector of triplets is used to represent a case. The elements of this vector describe the property, its importance (weight) within this case, and its value. The case e and its elements $\tau_i (i = 1, 2, \ldots, k)$ are defined as follows:

$$e = \{\tau_0, \tau_1, \ldots, \tau_k\}$$
$$\tau_i = (P_i, w_i, \chi_i)$$

where P_i is the property name, w_i its weight, and χ_i the value assigned to this property. This representation is an augmentation of the representation commonly used, with the addition of allowing each case to be represented by a distinct set of properties that are weighted independently. The domain of the property values is left unspecified. It is sufficient that there exists a similarity operator that will assess the similarity between two values for the same property.

Step 2: Similarity between fuzzy and nonfuzzy feature values

(a) *Similarity measure between two quantities in a specific range.* The crisp similarity measure between two quantities a and b in a specific range may be defined as

$$\text{SM}(a, b) = 1 - \frac{|b - a|}{\beta - \alpha}, \qquad a, b \in [\alpha, \beta] \tag{3.22}$$

where α and β are the lower and upper bounds of the range, respectively. Next we describe the concept of similarity between each of the types of values specified: crisp, range, and fuzzy.

(b) *Similarity between a crisp value and a range.* The augmentation of any similarity measure when comparing a crisp value a with a range $[b_1, b_2]$ is defined as

$$\text{SM}(a, [b_1, b_2]) = \frac{\int_{b_1}^{b_2} \text{SM}(a, x) \, dx}{b_2 - b_1} \tag{3.23}$$

(c) *Similarity between a crisp value and a fuzzy value.* If we represent a fuzzy value by the membership function $\mu_A(x)$, the similarity measure between a crisp value and the fuzzy value is defined as

$$SM(a, \mu_A) = \frac{\int_\alpha^\beta \mu_A(x)SM(a, x)\,dx}{\int_\alpha^\beta \mu_A(x)\,dx} \max_{x \in [\alpha, \beta]}\{\mu_A(x)\} \qquad (3.24)$$

The last multiplier is meant to compensate in situations where the membership function is not normalized. If the membership functions for all the fuzzy variables are normalized, the last multiplier can be omitted.

(d) *Similarity between a range and a fuzzy value.* The similarity between a given range $[a_1, a_2]$ and a fuzzy value, denoted by the membership function $\mu_A(x)$, is defined as

$$SM([a_1, a_2], \mu_A) = \frac{\int_{a_1}^{a_2} \int_\alpha^\beta \mu_A(y)SM(x, y)\,dy\,dx}{(a_2 - a_1)\int_\alpha^\beta \mu_A(x)\,dx} \max_{x \in [\alpha, \beta]}\{\mu_A(x)\} \qquad (3.25)$$

where a_1 and a_2 are in the range $[\alpha, \beta]$. Again, the final multiplier is used to compensate for nonnormalized membership functions. We represent a range as a fuzzy value whose membership function is equal to 1 within the range and zero outside it.

(e) *Similarity between two fuzzy values.* The similarity measure between two fuzzy values, represented by membership functions $\mu_A(x)$ and $\mu_B(x)$, is defined as

$$SM(\mu_A, \mu_B) = \frac{\int_\alpha^\beta \mu_A(x)\mu_B(y)SM(x, y)\,dy\,dx}{\int_\alpha^\beta \mu_A(x)\,dx \int_\alpha^\beta \mu_B(x)\,dx} \max_{x \in [\alpha, \beta]}\{\mu_A(x)\}$$
$$\times \max_{x \in [\alpha, \beta]}\{\mu_B(x)\} \qquad (3.26)$$

where $\mu_A(x)$ and $\mu_B(x)$ are the membership functions of the fuzzy values being compared. Note that by viewing ranges as fuzzy values whose membership grade is one within the range and zero outside it, this equation defines the similarity between two ranges.

Step 3: *General procedure for defining a family of similarity assessments*

(a) *Create an intermediate case based on the compared cases.* Given two cases—e, described by a set of properties P, and e', described by a set of properties P'—we define an intermediate case e_q that has the same number of elements as $P \cup P'$:

$$e_q = \{\tau_0, \tau_1, \dots, \tau_m\}$$
$$\Pi = P \cup P'$$

where τ_i denotes intermediate elements (i.e., each represents a triplet of property, weight, and value) used during the similarity calculation. The set Π contains the properties describing case e_q, and m is the cardinality of Π.

(b) *Compute the elements (triplets) of the intermediate case.* Associate a property with each element τ_i in the intermediate case, and calculate the weight and value for that property using those from the original cases that have the same property, in the following way:

$$\tau_i = (\Pi_i, w_i, \sigma_i)$$
$$\Pi_i = P_j = P'_k$$
$$w_i = \text{weight\_combination}(w_j, w'_k)$$
$$\sigma_i = \text{similarity\_operator}(\chi_j, \chi'_k)$$

where Π_i is one of the properties, w_i a combined weighting coefficient, and σ_i a similarity value (in the range $[0, 1]$). The triplet (P_j, w_j, χ_j) is an element of case e that has property $P_j = \Pi_i$. If there is no such element, w_i is set to a weight combination $(0, w'_k)$ and σ_i is set to zero. The same holds true for triplet (P'_k, w'_k, χ'_k) from case e'. The similarity operator is used to compute the similarity between the two original values, and it depends on the method of representation used for the values.

(c) *Compute the outcome similarity measure using the intermediate case.* After calculating each of the elements in the intermediate case, the weights are normalized (i.e., making their sum equal to 1), as shown below for $\overline{w}_i$ below. The similarity between the original cases $[\text{SM}(e, e')]$ is computed from the weighted sum of the discrete similarity values:

$$\overline{w}_i = \frac{w_i}{\sum_j w_j}$$
$$\text{SM}(e, e') = \sum_i \overline{w}_i \sigma_i$$

The weight combination is critically important to the quality of the similarity assessment, as is the similarity operator. There are several weight combination methods. These methods alter the weights of the similarity values for the various properties of the intermediate cases but not the values of the intermediate cases or the comparison procedure itself. Some of them are described here.

(1) *Minimum combination.* The nonnormalized weighting coefficient is taken as the minimum of the two weighting coefficients of the cases compared:

$$w_i = \min(w_j, w'_k)$$

This weight combination method is optimistic in the sense that properties appearing in only one of the cases will not affect the similarity measure.

This method tends to give less importance to properties that strongly affect only one of the cases. Thus, the similarity measure will be based on the "common denominator" of the two cases.

(2) *Maximum combination.* The nonnormalized weighting coefficient is taken as the maximum of the two weighting coefficients:

$$w_i = \max(w_j, w_k')$$

This weight combination method stresses the properties that are important in both cases. To achieve a high degree of similarity, the properties that are important to each of the cases must hold similar values in both of them.

(3) *Mean combination.* The nonnormalized weighting coefficient is taken as the mean of the two weighting coefficients. Note that since a normalization stage is conducted after this stage, a simple sum of the two coefficients is sufficient:

$$w_i = w_j + w_k'$$

This averaging weight combination method lowers the importance of properties that are important in only one of the compared cases.

(4) *Geometric mean combination.* The nonnormalized weighting coefficient is taken as the square root of the product of the two weighting coefficients:

$$w_i = \sqrt{w_j w_k'}$$

Like the minimum combination, this weight combination method is optimistic. Singular properties that appear in only one of the cases are not considered in the similarity calculation. For properties that are particularly irrelevant in one of the cases, this averaging approach allocates less importance to them than the "mean" combination.

(5) *Perpendicular combination.* The nonnormalized weighting coefficient is taken as the square root of the sum of the squares of the two weighting coefficients:

$$w_i = \sqrt{w_j^2 + w_k'^2}$$

This method reinforces properties that have high weight coefficients. It is a compromise, somewhere between the "mean" and "maximum" combination methods.

(6) *l-Power combination.* The nonnormalized weighting coefficient is taken as the *l*th root of the sum of the *l*th power of the two weighting coefficients:

$$w_i = \sqrt[l]{w_j^l + w_k'^l}$$

This weight combination method is a generalization of the perpendicular combination method. For values of l greater than 1, this method reinforces properties that have high weights. The higher the value of l is, the closer this combination method approaches the "maximum" combination (after normalization). For values of l smaller than 1, the lower the value of l is, the closer this method approaches the common denominator, the "minimum" combination (after normalization).

3.4 FUZZY CLASSIFICATION AND CLUSTERING OF CASES

In this section we describe how the similarity measures can be used for case matching and retrieval through classification or clustering of cases under supervised and unsupervised modes, respectively. In general, in the process of case matching and retrieval, the searching space is the entire case base, which not only makes the task costly and inefficient, but also sometimes leads to poor performance. To address such a problem, many classification and clustering algorithms are applied before selection of the most similar case or cases. After the cases are partitioned into several subclusters, the task of case matching and retrieval then boils down to matching the new case with one of the several subclusters, and finally, the desired number of similar cases can be obtained. Thus, various classification/clustering algorithms, such as fuzzy ID3 and fuzzy *c*-means, play an important role in this process.

To develop such classification/clustering algorithms, it is important to hypothesize that objects can be represented in an *n*-dimensional space where the axes represent the variables (features), objects (or entities) become points (vectors) in that space, and the clusters are compact groups of those points. The challenging part in this scheme is that the interpoint distances are dependent on the measurement scales used for the variables and metric chosen. Therefore, the concept of Euclidean space is important in constructing many similarity metrics.

Furthermore, in real-life problems, the clusters of cases (vectors) are usually overlapping, that is, their boundaries are fuzzy. The relevance of fuzzy logic in handling uncertainties arising from such overlapping classes has been addressed adequately [23,35–40]. As an example, here we describe a fuzzy ID3 algorithm and fuzzy *c*-means (FCM) algorithm, which are widely used for classification and clustering of cases, respectively. Fuzzy ID3 (fuzzy interactive dichotomizer 3) is an extension of the conventional ID3 algorithm, which is a decision tree based method using an information-theoretic approach. The main idea is to examine the feature that provides the greatest gain in information, equivalent, say, to the greatest decrease in entropy. Fuzzy ID3 can accept linguistic inputs, and the outputs are described in membership values, which are more reasonable and flexible in practice. Here one can receive a confidence factor while reaching a decision and determining the rule base. The detail procedure of the algorithm is given in Section 3.4.2. Fuzzy c-means algorithm (defined in Section 3.4.3) is a well-known clustering algorithm; the main idea is to minimize iteratively an appropriately defined objective function that reflects the proximities of the objects to the cluster centers.

Before we describe them, we define weighted intracluster and intercluster similarity measures, which may be used while partitioning a data set or evaluating the performance of a clustering/classification algorithm.

3.4.1 Weighted Intracluster and Intercluster Similarity

3.4.1.1 *Intracluster Similarity* For a given cluster LE, its intracluster similarity is defined as

$$SM_{LE}^{(w)} = \frac{2}{r(r-1)} \sum_{e_p, e_q \in LE(p<q)} SM_{pq}^{(w)} \tag{3.27}$$

where r is the number of cases in the cluster LE. For a partition containing m clusters $\{LE_1, LE_2, \ldots, LE_m\}$, the intracluster similarity may be defined as the average of all of its individual intracluster similarities:

$$SM_{int\,ra}^{(w)} = \frac{1}{m} \sum_{j=1}^{m} SM_{LE_j}^{(w)} \tag{3.28}$$

It is clear that the value of $SM_{int\,ra}^{(w)}$ lies in [0,1]. The higher the value of $SM_{int\,ra}^{(w)}$, the greater is the homogeneity of the clusters.

3.4.1.2 *Intercluster Similarity* For a pair of clusters LE_1 and LE_2, the intercluster similarity is defined as

$$SM_{LE_1,LE_2}^{(w)} = \frac{1}{r_1 r_2} \sum_{e_p \in LE_1, e_q \in LE_2} SM_{pq}^{(w)} \tag{3.29}$$

where r_1 and r_2 are the numbers of cases in LE_1 and LE_2, respectively. For a partition containing m clusters $\{LE_1, LE_2, \ldots, LE_m\}$, the intercluster similarity may be defined as the average of the intercluster similarities, computed over all pairs of constituting clusters:

$$SM_{int\,er}^{(w)} = \frac{2}{m(m-1)} \sum_{1 \le i < j \le m} SM_{LE_i,LE_j}^{(w)} \tag{3.30}$$

The value of $SM_{inter}^{(w)}$ lies in $[0,1]$. The smaller the value of $SM_{inter}^{(w)}$, the higher the separability between clusters.

Based on the foregoing measures, the following criterion is defined:

$$\min_{e_q \in LE_i} SM_{pq}^{(w)} > \max_{1 \leq j \leq m, j \neq i} \left(\max_{e_q \in LE_j} SM_{pq}^{(w)} \right) \qquad (3.31)$$

which may need to be satisfied for any $i(1 \leq i \leq m)$ and any case $e_p \in LE_i$ for generating m clusters. Inequality (3.31) means that the similarity between a case and the cluster to which it actually belongs is greater than the similarity between that case and any other cluster.

3.4.2   Fuzzy ID3 Algorithm for Classification

The fuzzy interactive dichotomizer 3 (Fuzzy ID3) algorithm that will be described here has been developed by Singal et al. [41]. Before we explain the algorithm we introduce, for the sake of convenience, the classical ID3 algorithm and the relevance of incorporating fuzzy sets into it.

3.4.2.1   *Conventional ID3*   This is a decision tree–based method that uses an information-theoretic approach. The procedure used in this technique is that at any point we examine the feature that provides the greatest gain in information or, equivalently, the greatest decrease in entropy. Entropy is defined as $-p \log_2 p$, where probability p is determined on the basis of the frequency of occurrence of that feature. In the general case, N labeled patterns are partitioned into sets of patterns belonging to classes $C_i, i = 1, 2, \ldots, l$. The population in class C_i is n_i. Each pattern has n features, and each feature has $j(\geq 2)$ values. The ID3 prescription for synthesizing an efficient decision tree can be stated as follows:

Step 1. Calculate the initial value of the entropy:

$$\text{Entropy}(I) = \sum_{i=1}^{l} -\frac{n_i}{N} \log_2 \frac{n_i}{N}$$

$$= \sum_{i=1}^{l} -p_i \log_2 p_i \qquad (3.32)$$

Step 2. Select the feature that results in the maximum decrease in entropy (i.e., a gain in information) to serve as the root node of the decision tree.

Step 3. Build the next level of the decision tree by selecting a feature that provides the next-greatest decrease in entropy.

Step 4. Repeat steps 1 through 3. Continue the procedure until all the subpopulations are of a single class and the system entropy is zero. At this stage one obtains a set of leaf nodes (subpopulations) of the decision tree, where the patterns are of a single class. Note that there may be some nodes that cannot be resolved further.

Note that conventional ID3 often fails to give any information when there are overlapping pattern classes. In this algorithm, we partition the sample space in the form of a tree by using attribute values only. When two sample points from two different overlapping classes lie in an intersecting region, the corresponding features for these samples are the same. This implies that they are associated with traveling through the same path in the decision tree, and finally, they arrive at the same node. We cannot split the node further because the gain in entropy would then be zero (always), which is one of the criteria used to stop building the tree. Thus, in the overlapped region, that attribute value fails to provide any decision about the relevant leaf node.

To obtain more information with respect to this problem, one needs to dig further into the data. Intuition tells us that the pattern points of any particular class must be clustered around some characteristic prototype or class center. We wish to exploit the fact that points nearer this center have a higher "degree of belonging" to that class than do points farther from it. This idea brings in the concept of fuzzy sets, which allows a pattern to have a finite nonzero membership to more than one class. Here lies the utility of employing fuzzy sets to model overlapping or ambiguous real-life pattern classes [23]. Moreover, the conventional ID3 algorithm can handle only discrete-valued or symbolic attributes. Real-life problems, on the other hand, require the modeling of continuous attributes. Fuzzy sets can also be useful in this regard.

3.4.2.2 *Fuzzy ID3*   Fuzzy set–theoretic concepts are introduced at the input, output, and node levels of the ID3 algorithm [35,42]. Linguistic inputs enable the handling of continuous attributes at the input. The output is evaluated in terms of class membership values. A fuzziness measure is computed at the node level to take care of overlapping classes. This is reducible to the classical entropy, used in the conventional ID3, in the crisp case. A confidence factor is estimated at the nodes while reaching a decision and determining the rule base.

Input Representation   An input feature value is described, as in Section 2.3.2, in terms of a combination of overlapping membership values in the linguistic property sets: low *(L)*, medium *(M)*, and high *(H)*. An n-dimensional pattern, $\hat{e}_i = [F_{i1}, F_{i2}, \ldots, F_{in}]$, is then represented as a $3n$-dimensional vector [17,42]:

$$\hat{e}_i = [\mu_{\text{low}(F_{i1})}(\hat{e}_i), \mu_{\text{medium}(F_{i1})}(\hat{e}_i), \mu_{\text{high}(F_{i1})}(\hat{e}_i), \ldots, \mu_{\text{high}(F_{in})}(\hat{e}_i)] \qquad (3.33)$$

where the μ values indicate the membership functions of the corresponding linguistic π-sets "low," "medium," and "high" along each feature axis. Each μ value is then discretized, using a threshold, to enable a convenient mapping in the ID3 framework.

When the input feature is numerical, the π-fuzzy sets are used whose one-dimensional form with range [0,1] is as follows [equation (2.9)]:

$$\pi(F_j; c, \lambda) = \begin{cases} 2\left(1 - \dfrac{\|F_j - c\|}{\lambda}\right)^2 & \text{for } \frac{\lambda}{2} \le \|F_j - c\| \le \lambda \\ 1 - 2\left(\dfrac{\|F_j - c\|}{\lambda}\right)^2 & \text{for } 0 \le \|F_j - c\| \le \frac{\lambda}{2} \\ 0 & \text{otherwise} \end{cases} \tag{3.34}$$

where $\lambda (> 0)$ is the radius of the π-function with c as the central point. Note that features in linguistic and set forms can also be handled in this framework [42].

Output Representation Consider an l-class problem domain. The membership of the ith pattern in class k, lying in the range [0,1], is defined as [23]

$$\mu_{ik}(\hat{e}_i) = \frac{1}{1 + (w_{ik}/f_d)^{f_e}} \tag{3.35}$$

where w_{ik} is the weighted distance of the training pattern $\hat{e}_i$ from class C_k, and the positive constants f_d and f_e are the denominational and exponential fuzzy generators controlling the amount of fuzziness in the class membership set.

Fuzziness Measure Fuzziness is incorporated at the node level by changing the decision function from classical entropy to a fuzzy measure, FM. This is defined as [41]

$$\begin{aligned} \text{FM}(I) &= \sum_{k=1}^{l} \left[\frac{1}{N} \sum_{i=1}^{N} \min(\mu_{ik}, \ 1 - \mu_{ik}) - \frac{n_k}{N} \log_2 \frac{n_k}{N} \right] \\ &= \sum_{k=1}^{l} \left[\frac{1}{N} \sum_{i=1}^{N} \min(\mu_{ik}, \ 1 - \mu_{ik}) - p_k \log_2 p_k \right] \end{aligned} \tag{3.36}$$

where N is the number of pattern points in the training set, l the number of classes, n_i the population in class C_i, p_k the a priori probability of the kth class, and μ_{ik} the membership of the ith pattern to the kth class. The expression for FM is defined so that when the class membership values are zero or 1 (crisp), it represents the classical entropy. The first term of equation (3.36) ensures that pattern points lying in overlapping regions are assigned lower weights during construction of the decision tree, which is an intuitively appealing idea. The reason for this lower weighting is that such ambiguous patterns (i.e., having μ values close to 0.5) lead to an increase in FM, thereby creating an obstacle to its minimization.

Estimating Confidence of the Nodes Mandal et al. [43] provided a scheme to calculate the *confidence factor* (CF) for a rule base to infer how much a point belongs to a particular class in terms of multiple choices (e.g., first choice, second choice). It is defined as

$$CF = \frac{1}{2}\left[\mu + \frac{1}{l-1}\sum_{j=1}^{l}(\mu - \mu_j)\right] \qquad (3.37)$$

where μ_j is the class membership of the jth pattern to class C_j and μ is the highest membership value. The concept of CF deals with the difficulty in assigning a particular class label, which depends not only on the highest entry μ but also on its differences from the other entries.

Note that when computing CF_1 and CF_2 (when a second choice is necessary), leaving out the first choice in equation (3.37) [43], μ is set equal to the first and second highest membership values, respectively. Let CF^k denote the CF_1 value of a pattern corresponding to class C_k (i.e., having the highest membership value). Then the rule base is as follows:

1. If $0.8 \leq CF^k \leq 1.0$, it is *very likely* to belong to class C_k, and there is no second choice.
2. If $0.6 \leq CF^k < 0.8$, it is *likely* to belong to class C_k, and there is second choice.
3. If $0.4 \leq CF^k < 0.6$, it is *more or less likely* to belong to class C_k, and there is second choice.
4. If $0.1 \leq CF^k < 0.4$, it is *not unlikely* to belong to class C_k, and there is no second choice.
5. If $CF^k < 0.1$, it is *unable to recognize* class C_k, and there is no second choice.

In the case of a single choice (when rule 1 fires), we update the confidence factor to 1, so that $CF_1 = 1$ and $CF_2 = 0$. For other choices (when rules 2 to 4 are in effect), we additionally compute CF_2, corresponding to the class with the second-highest membership value. Finally, an aggregation is made at the node level. Let us now describe the algorithm in detail.

Step 1. Calculate an initial value of the fuzziness measure (FM) using equation (3.36).

Step 2. Select a feature to serve as the root node of the decision tree.

(a) For each attribute $Attr_i, i = 1, 2, \ldots, 3n$, partition the original population into two subpartitions according to the values a_{ij} (where $j = 0$ or 1 and denotes the attribute value 0 or 1) of the attribute $Attr_i$. Although there are n_{ij} patterns associated with traveling down branch a_{ij}, these patterns need not necessarily belong to any single class.

(b) Evaluate FM for each branch.

(c) The decrease in FM, as a result of testing attribute $Attr_i$, is $\Delta FM(i) = FM(i) - FM(i, Attr_i)$.

(d) Select an attribute $Attr_k$ that yields the greatest decrease in FM for which $\Delta FM(k) > \Delta FM(i)$ for all $i = 1, 2, \ldots, l$, and $i \neq k$.

(e) The attribute $Attr_k$ is then the root of the decision tree.

Step 3. Build the next level of the decision tree. Select an attribute $Attr_{k'}$ to serve as the level 1 node, so that after testing on $Attr_{k'}$ along all branches, we obtain the maximum decrease in FM.

Step 4. Repeat steps 3 through 5. Continue the process until all subpopulations reaching a leaf node contain a single class or the decrease in FM, ΔFM, is zero. Mark the terminal nodes that have pattern points belonging to more than one class. Such nodes are called *unresolved nodes*.

Step 5. For each unresolved node, do the following.

(a) Calculate the confidence factors CF_1 and CF_2 as in equation (3.37).

(b) Identify the classes that have at least one CF_1 or CF_2.

(c) For each pattern point in the node, if $CF_1 \geq 0.8$, put $CF_1 = 1$, and $CF_2 = 0$.

(d) Consider the classwise average summation of the CF values.

(e) Mark the classes that get the highest and second-highest CF values. Declare this node to be the representative of the two relevant classes, with membership corresponding to the two highest CF values.

3.4.3 Fuzzy *c*-Means Algorithm for Clustering

Fuzzy *c*-means algorithm (FCM) was proposed by Dunn [44] and generalized by Bezdek [45]. In this approach and most of its extensions, the basic idea of determining the fuzzy clusters is by minimizing an appropriately defined objective function. The membership functions are based on a distance function, so that degrees of membership express proximities of data entities to the cluster centers. By choosing suitable distance functions, different cluster shapes can be identified. The algorithm is based on iterative minimization of the following objective function [45,46]:

$$J(U, V) = \sum_{i=1}^{c} \sum_{k=1}^{n} u_{ik}^{m} |x_k - v_i|^2 \tag{3.38}$$

where $x_1, x_2, \ldots, x_n$ are feature vectors; $V = \{v_1, v_2, \ldots, v_c\}$ are cluster centers; $m \in [1, \infty]$ and $U = [u_{ik}]$ is a $c \times n$ matrix, where u_{ik} is the ith membership value of the kth input sample x_k; and the membership values satisfy the following conditions:

$$0 \leq u_{ik} \leq 1, \qquad i = 1, 2, \ldots, c; \quad k = 1, 2, \ldots, n \tag{3.39}$$

$$\sum_{i=1}^{c} u_{ik} = 1, \qquad k = 1, 2, \ldots, n \tag{3.40}$$

$$0 < \sum_{k=1}^{n} u_{ik} < n, \qquad i = 1, 2, \ldots, c \tag{3.41}$$

In the objective function (3.38), Euclidean distances between each input case and its corresponding cluster center are weighted by the fuzzy membership values. The algorithm is iterative and uses the following equations:

$$v_i = \frac{1}{\sum_{k=1}^{n} u_{ik}^m} \sum_{k=1}^{n} u_{ik}^m x_{ik}, \qquad i = 1, 2, \ldots, c \qquad (3.42)$$

$$u_{ik} = \frac{\left[1/|x_k - v_i|^2\right]^{1/(m-1)}}{\sum_{j=1}^{c} \left[1/|x_k - v_j|^2\right]^{1/(m-1)}}, \qquad i = 1, 2, \ldots, c; \quad k = 1, 2, \ldots, n \qquad (3.43)$$

The fuzzy c-means clustering procedure consists of the following steps [46]:

Step 1. Initialize $U^{(0)}$ randomly, or based on an approximation; initialize $V^{(0)}$ and calculate $U^{(0)}$. Set the iteration counter $\alpha = 1$. Select the number of class centers c, and choose the exponent weight m.

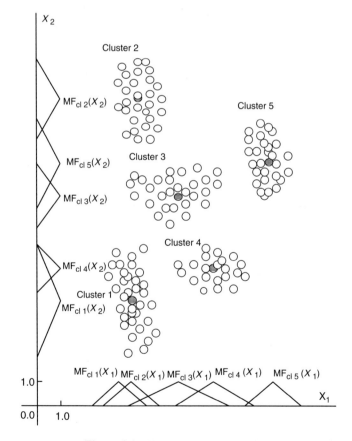

Figure 3.1 Fuzzy c-means clustering.

Step 2. Compute the cluster centers. Given $U^{(\alpha)}$, calculate $V^{(\alpha)}$ according to equation (3.42).

Step 3. Update the membership values. Given $V^{(\alpha)}$, calculate $U^{(\alpha)}$ according to equation (3.43).

Step 4. Stop the iteration if

$$\max \left| u_{ik}^{(\alpha)} - u_{ik}^{(\alpha-1)} \right| \leq \varepsilon \qquad (3.44)$$

else let $\alpha = \alpha + 1$ and go to step 2, where ε is a prespecified small number representing the smallest acceptable change in U.

Figure 3.1 illustrates the membership functions generated for a problem with two inputs (X_1 and X_2) and five clusters ($c = 5$).

3.5 CASE FEATURE WEIGHTING

In Sections 3.3 and 3.4 we have briefly described some similarity measures and then introduced the concept of fuzzy similarity, which are closely based on the feature weights. Essentially, feature weights describe the significance, and hence the relation, of various features. This means that if a feature is more relevant in terms of characterizing cases, it should have more weight (importance). Thus, case feature weighting is crucial in computing similarity, especially weighted distance and weighted feature similarity, which have been introduced in Section 3.4. The problem of weighting features (i.e., assigning weights to features) in a soft computing framework has been investigated by many researchers; see, for example, the recent unsupervised feature selection [47,48] and feature extraction [49] algorithms using a neuro-fuzzy approach [23] and a GA-based approach [50]. In this section, some of them are presented. Let us consider a CBR system where a case is usually represented by a feature vector with n components. Then the global feature weight refers to a *vector* $(w_1, w_2, \ldots, w_n)$, where each of its components is a real number in [0,1]. It can be interpreted in the way that for the entire case library, different features have different degrees of importance to the solution. This is different from the concept of a local feature weight. A global feature weight is assigned to each feature of a case, and the same feature for all the cases has the same global weight. On the other hand, a local feature weight is assigned to each feature of a specific case, and a feature may have different local weights for different cases. In this section, a gradient-based optimization technique and neural networks are used to learn the global feature weights.

Gradient-based optimization techniques are useful in determining the search directions based on the derivative information of an objective function. The optimization is carried out by minimizing a real-valued objective function defined on an n-dimensional input space. In learning the feature weights, an individual weight could be considered as an adjustable parameter of the objective function, which is usually nonlinear in form. The *method of steepest descent*, also known as *gradient descent*, is one of the earliest methods proposed for minimizing a (given) function defined on a multidimensional input space. We describe briefly how this method is being used in learning the global feature weights of a similarity function, such as equation (3.1).

Using the concept of weighted Euclidean distance, stated in Section 3.1, a feature evaluation function is defined (in which the feature weights are regarded as the variables). The smaller the value of the evaluation function, the better the corresponding feature weights are in describing all the cases. Thus, the purpose is to find weights that permit the evaluation function to attain its minimum. The task of minimizing the evaluation function with respect to the weights can be performed using any optimization method, such as a gradient descent technique, neural approach, and genetic algorithm–based approach.

When we do not have the necessary functional derivative information to search for a set of suitable weights, we may evaluate only the objective function and determine the subsequent search direction after each evaluation. The heuristic guidelines used to determine the search direction are usually based on simple intuitive concepts such as the idea of evolution. The major advantage of this type of derivative-free optimization technique is that it does not require the objective function to be differentiable, and the relaxation of this requirement allows us to develop complex objective functions. However, there is one major drawback to this type of technique—it cannot be studied analytically because of its random and problem-specific behaviors. One major representative of these techniques is the genetic algorithm (GA).

3.5.1 Using the Gradient Descent Technique and Neural Networks

For a given collection of feature weights w_j $(w_j \in [0,1], j = 1, 2, \ldots, n)$, and a pair of cases e_p and e_q, equation (3.1) defines a weighted distance measure $d_{pq}^{(w)}$ and equation (3.2) defines a similarity measure $\text{SM}_{pq}^{(w)}$. When all the weights take the value 1, $d_{pq}^{(w)}$ degenerates to the Euclidean distance $d_{pq}^{(1)}$, and $\text{SM}_{pq}^{(w)}$ to $\text{SM}_{pq}^{(1)}$.

A feature evaluation index function E corresponding to a set of weights w is defined as [51]

$$E(w) = \sum_{p} \sum_{q(q \neq p)} \left[\text{SM}_{pq}^{(w)}(1 - \text{SM}_{pq}^{(1)}) + \text{SM}_{pq}^{(1)}(1 - \text{SM}_{pq}^{(w)}) \right] \qquad (3.45)$$

It has the following characteristics:

- If $\text{SM}_{pq}^{(1)} = \text{SM}_{pq}^{(w)} = 0$ or 1, the contribution of the pair of cases to the evaluation index $E(w)$ is minimum.
- If $\text{SM}_{pq}^{(1)} = \text{SM}_{pq}^{(w)} = 0.5$, the contribution of the pair of cases to $E(w)$ becomes maximum.

- For $SM_{pq}^{(1)} < 0.5$ as $SM_{pq}^{(w)} \to 0$, $E(w)$ decreases.
- For $SM_{pq}^{(1)} > 0.5$ as $SM_{pq}^{(w)} \to 1$, $E(w)$ decreases.

Therefore, the feature evaluation index decreases as the similarity between the pth and qth cases in the transformed (weighted) feature space tends to either 0 (when $SM_{pq}^{(1)} < 0.5$) or 1 (when $SM_{pq}^{(1)} > 0.5$). In other words, the feature evaluation index decreases as the decision on the similarity between a pair of cases (i.e., whether they lie in the same cluster or not) becomes more and more crisp. This means that if the intercluster/intracluster distances in the transformed space increase/decrease, the feature evaluation index corresponding to the set of weights w decreases. Therefore, our objective is to extract those feature weights for which the evaluation index becomes minimum; thereby optimizing the decision on the similarity of a pair of patterns with respect to their belonging to a cluster.

The foregoing characteristics can be verified as follows. From equation (3.45) we have

$$\frac{\partial E(w)}{\partial SM_{pq}^{(w)}} = 1 - 2SM_{pq}^{(1)} \tag{3.46}$$

For $SM_{pq}^{(1)} < 0.5$, $(\partial E(w)/\partial SM_{pq}^{(w)}) > 0$. This signifies that $E(w)$ decreases (increases) with decrease (increase) in $SM_{pq}^{(w)}$. For $SM_{pq}^{(1)} > 0.5$, $(\partial E(w)/\partial SM_{pq}^{(w)}) < 0$. This signifies that $E(w)$ decreases (increases) with increase (decrease) in $SM_{pq}^{(w)}$. Since $SM_{pq}^{(w)} \in [0, 1]$, $E(w)$ decreases (increases) as $SM_{pq}^{(w)} \to 0(1)$ in the former case and $SM_{pq}^{(w)} \to 1(0)$ in the latter.

Suppose that a gradient descent technique is used for minimizing equation (3.45). Then the change in w_j (denoted by Δw_j) is computed as

$$\Delta w_j = -\lambda \frac{\partial E}{\partial w_j} \tag{3.47}$$

for $j = 1, 2, \ldots, n$, where λ is the learning rate.

For the computation of $\partial E/\partial w_j$, the following expressions are used:

$$\frac{\partial E(w)}{\partial w_j} = \sum_p \sum_{q(q<p)} \left(1 - 2 \cdot SM_{pq}^{(w)}\right) \frac{\partial SM_{pq}^{(w)}}{\partial d_{pq}^{(w)}} \frac{\partial d_{pq}^{(w)}}{\partial w_j} \tag{3.48}$$

$$\frac{\partial SM_{pq}^{(w)}}{\partial d_{pq}^{(w)}} = \frac{-\alpha}{(1 + \alpha d_{pq}^{(w)})^2} \tag{3.49}$$

$$\frac{\partial d_{pq}^{(w)}}{\partial w_j} = \frac{w_j \chi_j^2}{\left(\sum_{j=1}^n w_j^2 \chi_j^2\right)^{1/2}} \qquad \text{where } \chi_j^2 = (x_{pj} - x_{qj})^2 \tag{3.50}$$

To obtain the optimal global feature weights, we need to minimize E, and the training algorithm used is described as follows:

Step 1. Select the parameter α and the learning rate λ.

Step 2. Initialize w_j with random values in [0,1].

Step 3. Compute Δw_j for each j using equation (3.47).

Step 4. Update w_j with $w_j + \Delta w_j$ for each j.

Step 5. Repeat steps 3 and 4 until convergence, that is, until the value of E becomes less than or equal to a given threshold (Th, say), or until the number of iterations exceeds a certain predefined number.

After training, the function $E(w)$ attains a local minimum. It is expected that on average, similarity values with trained weights $\{SM_{pq}^{(w)}, p, q = 1, 2, \ldots, N, q < p\}$, are closer to 0 or 1 than those without trained weights, such as $\{SM_{pq}^{(1)}, p, q = 1, 2, \ldots, N, q < p\}$. Note that $SM_{pq}^{(w)}$ [equation (3.2)] does not require the information on class label of cases, and therefore the global weight learning method is unsupervised. Moreover, this unsupervised method performs the task of feature weighting without clustering the feature space explicitly and does not need to know the number of clusters present in the feature space.

Recently, Pal et al. [51] have designed a connectionist framework to implement this learning strategy for determining weights automatically based on the searching principle of gradient descent technique. Figure 3.2 shows the network model, which consists of input, hidden, and output layers. The input layer consists of a pair of nodes corresponding to each feature; there are $2n$ nodes in the input layer, for n-dimensional (original) feature space. The hidden layer consists of n nodes, which

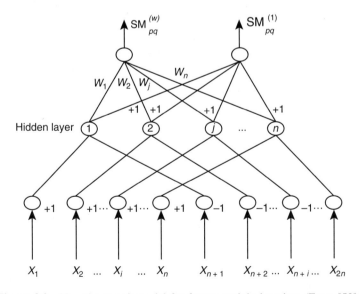

Figure 3.2   Neural network model for feature weight learning. (From [50].)

compute part χ_i^2 of equation (3.1). The output layer consists of two nodes, one of which computes $SM_{pq}^{(1)}$, and the other, $SM_{pq}^{(w)}$. The index $E(w)$ [equation (3.45)] is computed from these SM_{pq} values of the network.

Input nodes receive activations corresponding to feature values of each pair of patterns. A jth hidden node is connected only to an ith and $(i + n)$th input nodes via weights $+1$ and -1, respectively, where $j,i = 1,2,\ldots,n$ and $j = i$. The output node computing $SM_{pq}^{(w)}$ values is connected to a jth hidden node via weight $W_j(= w_j^2)$, whereas that computing $SM_{pq}^{(1)}$ values is connected to all the hidden nodes via weights of $+1$ each. During learning, each pair of patterns is presented at the input layer and the evaluation index is computed. The weights W_j are updated using the gradient descent technique in order to minimize the index $E(w)$. After minimization, such as, when $E(w)$ attains a local minimum, the weights $W_j(= w_j^2)$ of the links connecting hidden nodes and the output node computing $SM_{pq}^{(w)}$ values indicate the order of importance of the features. Note that this unsupervised method performs the task of feature selection without clustering the feature space explicitly and does not need to know the number of clusters present in the feature space.

Although we have described the foregoing methodology in unsupervised mode, a similar learning algorithm in a supervised framework for CBR problems can be derived based on the feature selection strategy reported [52].

3.5.2   Using Genetic Algorithms

Genetic algorithms [53,54] are derivative-free stochastic optimization methods based loosely on the concepts of natural selection and evolutionary processes [55]. (See Appendix C for basic principles and operations of GAs.) Their popularity can be attributed to their freedom from dependence on functional derivatives and to their incorporation of the following characteristics:

- GAs are parallel-search procedures that can be implemented on parallel processing machines for a massive increase in their operational efficiency.
- GAs are applicable to both continuous and discrete (combinatorial) optimization problems.
- GAs are stochastic and less likely to get trapped in local minima, which are inevitably present in any practical optimization application.
- GAs offer flexibility that facilitates both structure and parameter identification in complex models such as neural networks and fuzzy inference systems.

Dubitzky and Azuaje [50] developed an evolutionary computing approach for learning of global and local feature weights using the CBR system architecture depicted in Figure 3.3. The retrieval of cases is based on partitioning the cases into subsets. The part of the diagram above the dashed line in Figures 3.3a and b demonstrates how supervised and unsupervised learning interacts with the system components. Typical supervised learning methods include kd-trees, shared-feature and discrimination networks, decision trees, and artificial neural networks, and unsupervised methods include techniques such as automatic cluster detection.

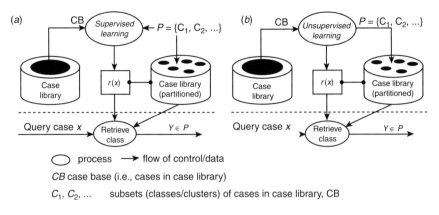

Figure 3.3 Case retrieval based on partitioning. (From [50].)

The knowledge structure [i.e., $r(e)$] that is learned may be in the form of a set of rules or a decision tree, which effectively selects a class or cluster of cases, $Y \in P = \{C_1, C_2, C_3, \ldots, C_n\}$, where P denotes a partition, from the case library (CB) based on the query case, e.

Determination of the case feature weights is carried out by an introspective learning process. Figure 3.4 illustrates the basic structure used in finding the feature weights. The entire set of available cases is divided randomly into three sets: training, testing set, and evaluation sets. In each learning cycle, the newly determined feature weights are used to test the performance of the case base using test cases. If a satisfactory threshold is reached, the case base is evaluated against a set of testing cases. Figure 3.4 shows the architecture for learning case feature weights through a combined introspective and evolutionary learning approach. This framework can be used to learn both the global and local weights. The use of genetic operators is shown in the box labeled "adjust weights." For determination of both global and local case feature weights, a chromosome was represented by a vector $\mathbf{Sr}$ of n 10-digit binary bit strings $Sr_j, j = 1, 2, \ldots, n$, [i.e., $\mathbf{Sr} = (Sr_1, Sr_2, \ldots, Sr_n)$], where n is the number of features used to describe a case. Each bit string, Sr_j, represents the weight w_j of a case feature.

One of the notable things in the algorithm is that the mutation operation on the feature weight vector chromosome $\mathbf{Sr}$ was divided into macro- and micromutation operations. A random selection scheme determines which one needs to be used when mutation is necessary. Micromutation randomly selects a single macrogene Sr_j from $\mathbf{Sr}$ and modifies (mutates) it by altering a single bit randomly. On the other hand, macromutation randomly selects a single macrogene Sr_j from $\mathbf{Sr}$ and replaces it entirely by a new randomly generated macrogene.

Although the same genetic operations were used to determine both global and local feature weights, the main differences between the two approaches lie mainly in the following aspects:

- Added complexity arises in the case of the local weight model, because for each of the cases in the training set (case base) a population of m

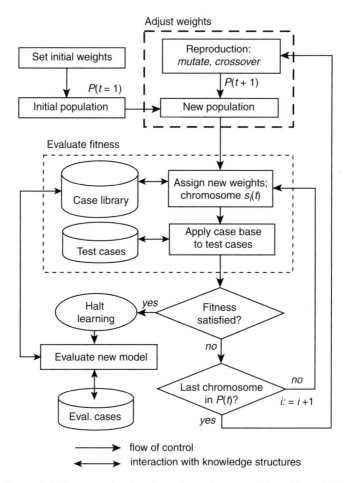

Figure 3.4   Introspective learning of case feature weights. (From [50].)

chromosomes (i.e., m weight sets) needs to be established, manipulated, and evaluated, in contrary to only m global weights in the global weight model.

- For the global weight model, it is straightforward to assign to the training cases m times the weights corresponding to the chromosomes of each generation and determine the overall error (or fitness), whereas for the local weight model it is not so obvious how to measure the fitness.

Dubitzky and Azuaje implemented the aforesaid scheme to predict the change in coronary heart disease (CHD) risk over a certain period of time. CHD is a degenerative disease that is a result of an increase of atheroma (degeneration of artery walls caused by the formation of fatty plaques or scar tissue) in coronary artery walls, leading to total or partial occlusion (blockage). The resulting clinical feature is myocardial infarction (heart attack) and subsequently, death. The case features considered include age (years), total cholesterol, HDL (high-density lipoprotein cholesterol), cigarette smoking (years), and SBP (systolic blood pressure). The performance of both learning models, within the context of risk-change prognosis, was compared with the following three reference models:

- A CBR model with global feature weight settings that were obtained from a participating domain expert
- A neural network model based on a standard back-propagation approach
- A standard multiple linear regression statistical model

The results indicated that the global weight learning approach is significantly superior to other models in terms of both prediction accuracy and learning time. Details of the methodology are available in Dubitzky and Azuaje [50].

3.6 CASE SELECTION AND RETRIEVAL USING NEURAL NETWORKS

As discussed before, case retrieval essentially means matching of patterns, where a current input pattern (case) is matched with one or more stored patterns or cases. Artificial neural networks (ANNs) are very efficient for performing this task, particularly when the data are incomplete, noisy, or imprecise. For the last few years, attempts have been made to develop methodologies that integrate case-based reasoning and ANNs. The problems associated with this integration include designing hybrid case-based connectionist systems [56–58], formulating connectionist indexing approaches [59,60], retrieval of cases using a neural network [61,62], and learning of cases in a connectionist framework [63]. Basic ANN models and their characteristics are described in Appendix B for the convenience of readers.

Using neural networks for case retrieval presents some problems of its own. For example, it is impractical for a single neural network to be expected to retrieve a single case from hundreds of possible classes of cases. The size of the network would then be too big, and consequently, the retrieval process would be too slow. Therefore, it may be desirable that such a large number of cases be grouped into similar types: first, grouped at a high level, and then each of those high-level groups grouped into further and more detailed subgroups. This means that one neural network can be used to classify all the categories of cases into the highest-level classes. For each of those high-level classes, another network can be used to divide the cases from that class into its subclasses. This may be repeated for each subclass until the lowest-level subclasses are reached.

Another way to handle this problem is to use the concept of *modularity* based on the principle of "divide and conquer." Here the main problem is split into simpler subproblems, each subproblem is then dealt with in a simpler network, and finally, the subnetworks are concatenated to determine the final network [64,65]. This approach provides gain in performance and reduction in learning time and network size. It also helps in preserving the identity of individual classes during training and in reducing the catastrophic interference due to the patterns from overlapping classes.

Again, domains that use case-based reasoning technique are usually complex. Therefore, a multilayered network is required [66] for handling the issue of nonlinearity in classification of cases at each level, or in dealing with a subproblem. If it is known how each case is classified into classes and subclasses, or each problem is split into subproblems, the learning task involved is supervised; otherwise, it is unsupervised. It may be mentioned here that an unsupervised neural learning algorithm for feature weight selection has been described in Section 3.5. In the following section we illustrate a supervised scheme for learning feature weights using neural networks for case selection and retrieval.

3.6.1 Methodology

Let $CB = \{e_1, e_2, \ldots, e_N\}$ denote a case library, $\{F_j \ (j = 1, 2, \ldots, n)\}$ denote a collection of features, and a variable V represent the action. The ith case e_i in the library can be represented as an $(n + 1)$-dimensional vector, $e_i = (x_{i1}, x_{i2}, \ldots, x_{in}, \theta_i)$, where x_{ij} corresponds to the value of feature $F_j \ (1 \leq j \leq n)$ and θ_i corresponds to the value of action $V(i = 1, 2, \ldots, N)$. The actions $\theta_i (i = 1, 2, \ldots, N)$ are considered to be class symbols, and the total number of classes is M. The M classes are denoted by $CM = \{C_1, C_2, \ldots, C_M\}$.

- *Input layer.* The number of nodes in this layer is equal to the number of features (n) of the case base. Each node represents a feature.
- *Hidden layer.* The number of nodes (l) in this layer is determined depending on the application domain. Experimentally, the number is greater than n but less than $2n$.
- *Output layer.* This contains M nodes, where M is the number of classes or clusters. Each node represents a fuzzy cluster (i.e., a discrete fuzzy set defined on the cluster space CM). The output value of each node represents the membership value, indicating the degree to which the training case belongs to the cluster corresponding to the node.

The popularly used sigmoid function is selected as the activation function. For a given input case (say, the mth case, e_m, $1 \leq m \leq N$), the propagation process of the input vector is described as follows:

$$\text{Input layer:} \ \{x_{mi} \mid i = 1, 2, \ldots, n\} \ \text{(of the given input vector)}$$

$$\text{Hidden layer:} \ y_{mj} = f\left(\sum_{i=1}^{n} u_{ij} x_{mi}\right), \quad j = 1, 2, \ldots, l \quad (3.51)$$

$$\text{Output layer:} \ \mu_{mk} = f\left(\sum_{j=1}^{l} v_{jk} y_{mj}\right), \quad k = 1, 2, \ldots, M \quad (3.52)$$

Here u_{ij} and v_{jk} represent the connection weights of the neural network, and the notation f represents the sigmoid function defined as $f(x) = 1/(1 + e^{-x})$. This is a traditional full connection network with three layers. The standard BP algorithm can be used to train this network. In other words, the popular gradient descent technique can be used to find the values of weights u_{ij} and v_{jk} such that the error function

$$\text{Er} = \sum_{m=1}^{N} \left[\frac{1}{2} \sum_{k=1}^{M} (\mu_{mk} - \theta_{mk})^2 \right] = \sum_{m=1}^{N} \text{Er}_m \qquad (3.53)$$

achieves a local minimum, where θ_{mk} taking either 0 or 1 corresponds to the action of the mth case. For example, $(\theta_{m1}, \theta_{m2}, \ldots \theta_{mM}) = (1, 0, \ldots, 0)$ if the mth case belongs to the first cluster.

After the training is complete, a set of fuzzy membership values on the cluster space $\{C_1, C_2, \ldots, C_M\}$ can be given for each case according to equation (3.52). Denoting these values by $(\mu_{m1}, \mu_{m2}, \ldots, \mu_{mM})$, where each component μ_{mj} represents the degree to which the mth case belongs to the jth cluster, we can reclassify the case base according to the following criteria. Consider a case e_m with values $(\mu_{m1}, \mu_{m2}, \ldots, \mu_{mM})$ to be reclassified. Let α and β be two thresholds determined a priori.

- *Criterion A.* If $\text{Entropy}(e_m) < \beta$ and $\mu_{mk} = \max_{1 \leq j \leq M} \mu_{mj} \geq \alpha$, the case e_m is classified into the kth cluster, where

$$\text{Entropy}(e_m) = -\sum_{j=1}^{M} \mu_{mj} \log \mu_{mj} \qquad (3.54)$$

- *Criterion B.* If $\text{Nonspec}(e_m) < \beta$ and $\mu_{mk} = \max_{1 \leq j \leq M} \mu_{mj} \geq \alpha$, the case e_m is classified into the kth cluster, where

$$\text{Nonspec}(e_m) = \sum_{j=1}^{M} \left(\mu_{mj}^* - \mu_{m(j+1)}^* \right) \log \mu_{mj}^* \qquad (3.55)$$

in which $(\mu_{m1}^*, \mu_{m2}^*, \ldots, \mu_{mM}^*)$ is the permutation of $(\mu_{m1}, \mu_{m2}, \ldots, \mu_{mM})$, sorted so that $\mu_{mj}^* \geq \mu_{m(j+1)}^*$ for $j = 1, 2, \ldots, M$ and $\mu_{m(M+1)}^* = 0$.

Criterion A is based on the fuzzy entropy [equation (3.54)] [67], which will tend to zero when all μ_{mj} tend either to zero or 1. Criterion B is based on the nonspecificity [equation (3.55)] [68], which will tend to zero when only one μ_{mj} tends to 1 and the others tend to zero. The fuzzy entropy and the nonspecificity represent two different types of uncertainties. According to criterion A or B, the case base can be classified into $M + 1$ clusters. The $(M + 1)$th cluster, called the *odd* (*unclassified*) *class*, contains the cases that cannot be classified into one of the M classes. The cases in the odd class have poor training results; that is, the output μ values of the network corresponding to these patterns have more than one component with high value, thereby causing difficulty in assigning them to a particular class. Since these odd cases mainly come from the overlapping regions of the current feature space, their number can be reduced by learning the weights (importance) of the individual features so that a transformed feature space is generated for representing the cases with reduced overlapping regions. Such methods have been described in Section 3.4 using gradient descent technique, a neuro-fuzzy approach, and a GA-based approach.

After classification of the N cases in the case base is made into $M + 1$ classes, the task of retrieval becomes a problem of mapping a query case to one of these classes. There are many ways to select a set of representative cases from the class retrieved. A simple way is to use the concept of *case density* of a case e in its class C, which is defined as

$$\text{Case density}(e, C) = \frac{\sum_{e' \in C - \{e\}} \text{SM}_{ee'}^{(w)}}{|C| - 1} \tag{3.56}$$

where $\text{SM}_{ee'}^{(w)}$ represents the weighted similarity between cases e and $e'; e, e' \in C, e \neq e'$ as defined by equation (3.2). $|C|$ is the number of cases in class C. The task of selecting a set of prototype cases from class C is therefore to compute case densities for all its cases, ranking them in descending order, and selecting a desired set of cases having higher-density values.

3.6.2 Glass Identification

To illustrate the effectiveness of using a neural network for case selection and retrieval, we describe here some results on a glass identification problem [69]. The glass database available in UCL (University College London, London) Machine Learning Repository [70] consists of 214 records, and each record has 11 attributes:

1. Id number: 1 to 214
2. RI: refractive index
3. Na: sodium (unit measurement is weight percent in corresponding oxide; the same unit measure is applicable to attributes 4 to 10)
4. Mg: magnesium
5. Al: aluminum
6. Si: silicon
7. K: potassium
8. Ca: calcium
9. Ba: barium
10. Fe: iron
11. Type of glass: window glass/non-window glass

TABLE 3.1 Sample Cases

RI	Na	Mg	Al	Si	K	Ca	Ba	Fe	Type
1.51215	12.99	3.47	1.12	72.98	0.62	8.35	0	0.31	W
1.51768	12.56	3.52	1.43	73.15	0.57	8.54	0	0	W
1.51652	13.56	3.57	1.47	72.45	0.64	7.96	0	0	W
1.51969	12.64	0	1.65	73.75	0.38	11.53	0	0	N
1.51754	13.39	3.66	1.19	72.79	0.57	8.27	0	0.11	W
1.51911	13.9	3.73	1.18	72.12	0.06	8.89	0	0	W

Table 3.1 shows some typical data with feature values. The neural network package Nerualworks Professional II/Plus version 5.3 was used to investigate the performance of the back-propagation model for fuzzy classification of the data. After about 50,000 cycles of training, the network converged and the RMS (root-mean-square) error was 0.1322. The number of correctly classified samples in each category and the cases selected thereafter when $\alpha = 0.95$ are depicted in Table 3.2. The number of samples in the ODD (unclassified) class was found to be 5 and 8, corresponding to window and non-window glasses. Only cases with case density greater than 0.95 are selected here, thereby giving rise to a reduction of about 30% in the case library.

3.7 CASE SELECTION USING A NEURO-FUZZY MODEL

In the present section we describe the method of De and Pal [71] for addressing the problem of selecting cases from overlapping class regions in a neuro-fuzzy framework. While the learning capability and case representation aspect are taken care of by neural networks, fuzzy set theory handles the uncertainty arising due to overlapping cases. Here the cases are stored as network parameters. A notion of fuzzy similarity, using a π-type membership function, is incorporated together with repeated *insertion* and *deletion* of cases to determine a stable case base. The architecture of the network is determined adaptively through the *growing* and *pruning* of hidden nodes under supervised training. The effectiveness of the cases selected by the network is demonstrated for a pattern classification problem using the 1-NN rule, with the cases as the prototypes. Both results and comparisons are presented for various artificial and real-life data for different parameter values of the similarity function, which controls the number of cases.

TABLE 3.2 Case Selection Result

Class	Samples Correctly Classified	Cases Selected
Window glass	158	112
Non-window glass	43	29

3.7.1 Selection of Cases and Class Representation

The method described here selects a few samples from each class as representative cases. (For the sake of convenience, the samples that are not selected as cases are referred to as patterns in subsequent discussion.) Let $\hat{e} = [x_1, x_2, \ldots, x_i, \ldots, x_n]$ be a pattern vector of known classification in an n-dimensional feature space containing M classes. The variable $e_{l_k} = [x_{l_k 1}, x_{l_k 2}, \ldots, x_{l_k i}, \ldots, x_{l_k n}]$ denotes the l_kth case from the kth class C_k. $\mu_{l_k}(\hat{e})$ represents the degree of similarity of $\hat{e}$ to a case e_{l_k}. $d_{l_k}(\hat{e})$ stands for the distance between $\hat{e}$ and e_{l_k}. The degree of similarity between a pattern $\hat{e}$ and a case e_{l_k} is defined as

$$
\mu_{l_k}(\hat{e}) = \begin{cases} 1 - 2\left[\dfrac{d_{l_k}(\hat{e})}{\lambda}\right]^2 & 0 \le d_{l_k}(\hat{e}) < \frac{\lambda}{2} \\[2mm] 2\left[1 - \dfrac{d_{l_k}(\hat{e})}{\lambda}\right]^2 & \frac{\lambda}{2} \le d_{l_k}(\hat{e}) < \lambda \\[2mm] 0 & \text{otherwise} \end{cases} \tag{3.57}
$$

Here λ is the bandwidth of $\mu_{l_k}(\hat{e})$ [i.e., the separation between its two (crossover) points where $\mu_{l_k} = 0.5$]. Note that $\mu_{l_k}(\hat{e})$ can be viewed as a π-type membership function characterizing a fuzzy set of points representing a region Rg_{l_k} with e_{l_k} as its center [72]. The distance $d_{l_k}(\hat{e})$ may be expressed in many ways. Considering a Euclidean norm, we have

$$
d_{l_k}(\hat{e}) = \left[\sum_{i=1}^{n}(x_i - x_{l_k i})^2\right]^{1/2} \tag{3.58}
$$

It is clear from equation (3.57) that $\mu_{l_k}(\hat{e})$ decreases with an increase in $d_{l_k}(\hat{e})$, and vice versa. The value of $\mu_{l_k}(\hat{e})$ is maximum (= 1.0) when $d_{l_k}(\hat{e})$ is zero (i.e., if a pattern $\hat{e}$ and the l_kth case are identical). The value of $\mu_{l_k}(\hat{e})$ is minimum (= 0) when $d_{l_k}(\hat{e}) \ge \lambda$. When $d_{l_k}(\hat{e}) = \lambda/2 = 0.5$, an ambiguous situation arises.

A pattern $\hat{e}$ is selected randomly from any class C_k, and it is considered to be the first case if the case base (CB_k) corresponding to that class (C_k) is empty. Otherwise, the $\mu_{l_k}(\hat{e})$ values are computed that correspond to the cases e_{l_k} in the case base, CB_k. The pattern $\hat{e}$ is selected as a new case, if $\mu_{l_k}(\hat{e}) \le 0.5, \forall l_k$.

When a case is selected, it is *inserted* into the case base. After repeating this process over all of the training patterns, a set of cases for each class is obtained, and these constitute the case base. The case base, CB, for the entire training set is the union of all of the CB_k (i.e., $\text{CB} = \bigcup_{k=1}^{M} \text{CB}_k$). After the formation of this case base, CB, a case e_{l_k} for which $\mu_{l_k}(\hat{e}) \le 0.5$ is minimum is *deleted* from CB if there are a number of patterns with $\mu_{l_k}(\hat{e}) > 0.5$ [or with $d_{l_k}(\hat{e}) < \lambda/2$]. The processes of *insertion* and *deletion* are repeated until the case base becomes stable; that is, the set of cases does not show any further changes. The *deletion* process reduces the possibility that a spurious pattern will be considered to be a representative case.

Therefore, the class C_k can be viewed as a union of all the regions (Rg_{l_k}) around its different cases, that is,

$$C_k = \bigcup_{l_k=1}^{s_k} Rg_{l_k}$$

where s_k is the number of cases in class C_k. Note that as the value of λ increases, the extent of the regions (Rg_{l_k}) representing different areas around the cases (e_{l_k}) increases, and therefore the number of cases s_k decreases. This implies that the generalization capability of an individual case increases with an increase in λ. Initially, although the number of cases will decrease with an increase in λ, the generalization capability of individual cases dominates. If λ continues to increase, the number of cases becomes so low that the generalization capability of the individual cases may not cope with the proper representation of the class structures.

3.7.2 Formulation of the Network

The network architecture is determined adaptively through growing and pruning of the hidden nodes. These *growing* and *pruning* activities correspond to the respective tasks of *insertion* and *deletion* of cases.

3.7.2.1 Architecture   The connectionist model (see Fig. 3.5) consists of three layers: the input, hidden, and class layers. The input layer represents the set of input features, so that for each feature there is a node (called an *input node*) in the input layer. Similarly, for each case there is a node in the hidden layer. For each hidden node, there is an auxiliary node that makes the hidden node turn on or off. An auxiliary node sends a signal back to the input layer only when it sends a signal to the corresponding hidden node that makes it turn on. The hidden nodes are turned on one at a time, while the remaining nodes are kept off. To retain the class information of the cases, a class layer consisting of several nodes, where each node (class node) represents a class, is considered.

The input nodes are connected to the hidden and auxiliary nodes by feedforward and feedback links, respectively. The weight of a feedforward link connecting the ith input node with the l_kth hidden node is

$$w_{l_k i}^0 = 1, \quad \forall\, l_k \text{ and } i \tag{3.59}$$

The weight $w_{l_k i}^{(\text{fb})}$ of a feedback link connecting the auxiliary node (corresponding to the l_kth hidden node) with the ith input node is the same as the ith feature value of the l_kth case $(x_{l_k i})$. That is,

$$w_{l_k i}^{(\text{fb})} = \xi_{l_k i} \tag{3.60}$$

The hidden layer is connected to the class layer via feedforward links. The weight $(w_{kl_k}^{(1)})$ of the link connecting the l_kth hidden node with the kth class node is 1 if and

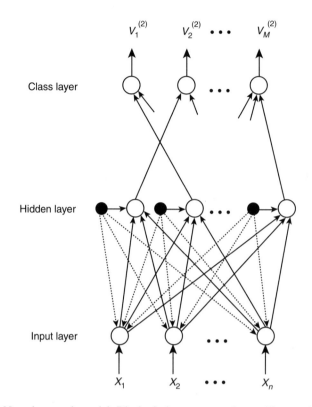

Figure 3.5 Neural network model. Black circles represent the auxiliary nodes, and white circles represent input, hidden, and class nodes. (From [71].)

only if the case corresponding to the hidden node belongs to class C_k. Otherwise, there is no such link between the hidden node and the class node. That is,

$$w_{kl_k}^{(1)} = \begin{cases} 1 & \text{if } e_{l_k} \in C_k \\ 0 & \text{otherwise} \end{cases} \tag{3.61}$$

At the beginning of the process, since the case base is empty, there are no hidden nodes. Hence, the connectivity between the layers is not established. When there is at least one hidden node, a pattern, $\hat{e}$, is presented to the input layer of the network. The activation of the ith input node when the l_kth hidden node is on is given by

$$v_{l_k i}^{(0)} = (u_{l_k i}^{(0)})^2 \tag{3.62}$$

$u_{l_k i}^{(0)}$ is the total input received by the ith input node when the l_kth hidden node is on, and this is given by

$$u_{l_k i}^{(0)} = x_i - u_{l_k i}^{(\text{fb})} \tag{3.63}$$

Here $u_{l_k i}^{(fb)} = (-1)w_{l_k i}^{(fb)}$ is the feedback input received by the input node. (The -1 is the feedback activation value of the auxiliary node corresponding to the l_kth hidden node.) The total input received by the l_kth hidden node when it is on is given by

$$u_{l_k}^{(1)} = \sum_i v_{l_k i}^{(0)} w_{l_k i}^{(0)} \tag{3.64}$$

The activation function of the l_kth hidden node is the same as $\mu_{l_k}(x)$ [i.e., equation (3.57)]. Thus, the activation $(v_{l_k}^{(1)})$ of the l_kth hidden node is given by

$$v_{l_k}^{(1)} = \begin{cases} 1 - 2\left[\dfrac{(u_{l_k}^{(1)})^{1/2}}{\lambda}\right]^2 & 0 \le (u_{l_k}^{(1)})^{1/2} < \frac{\lambda}{2} \\[2mm] 2\left[1 - \dfrac{(u_{l_k}^{(1)})^{1/2}}{\lambda}\right]^2 & \frac{\lambda}{2} \le (u_{l_k}^{(1)})^{1/2} < \lambda \\[2mm] 0 & \text{otherwise} \end{cases} \tag{3.65}$$

Here the value of λ is stored in all the hidden nodes.

3.7.2.2 *Training and Formation of the Network*

The network described in Section 3.7.2.1 is formed through the growing and pruning of the hidden nodes during the supervised training phase. Initially, there are only input and class layers. The patterns are presented in a random sequence to the input layer of the network. The first pattern presented to the network is considered as a case. A hidden node along with its auxiliary node representing this case is added to the network. The connections between these auxiliary and hidden nodes and the input and class layers are established as described by equations (3.59) to (3.61). For the remaining patterns, their degrees of similarity with the cases represented by the existing hidden node are computed, and if they are considered to be new cases (see Section 3.8.1), hidden nodes are added through growing operation. After the process of addition is over, it is checked if there is any redundant hidden node. This is done through a pruning operation depending on the criterion mentioned in Section 3.8.1. In this connection, one may note that as λ increases, the number of cases decreases, along with the number of hidden nodes. These two operations, which together constitute a single iteration, are continued until the structure of the network becomes stable; that is,

$$\sum_k \sum_{l_k i} \left|w_{l_k i}^{(fb)}(t)\right| = \sum_k \sum_{l_k i} \left|w_{l_k i}^{(fb)}(t-1)\right| \tag{3.66}$$

where t is the number of iterations. The aforesaid growing and pruning operations are described below.

Growing Hidden Nodes For a pattern $\hat{e} \in C_k$, if $v_{l_k}^{(1)} \le 0.5$ and $w_{l_k}^{(fb)} = e_{l_k} \in C_k$ for all the hidden nodes, x is added as a case. A hidden node along with its auxiliary node is added to the network to represent this case, and the links are established using equations (3.59) to (3.61). Note that the task of inserting cases, described in Section 3.8.1, is performed through this process.

Pruning Hidden Nodes   An l_k hidden node is deleted if

$$v_{l_k}^{(1)} = \min_{e_{l_k} = w_{l_k}^{(fb)} \in C_k} v_{l_k}^{(1)} \leq 0.5 \tag{3.67}$$

and the number of training samples for which $v_{l_k}^{(1)} > 0.5$ is less than a predefined value. Note that the task of deleting cases, described in Section 3.8.1, is performed through this process.

3.7.2.3 *1-NN Classification Using the Cases*   To demonstrate the effectiveness of the network model (i.e., the capability of the cases to represent their respective classes) for pattern classification, the principle of the 1-NN rule, with the cases as the prototypes, is considered. According to this rule, an unknown sample, $\hat{e}$, is said to be in class C_j for an L_jth case if

$$v_{L_j}^{(1)} = \max_{k,l_k} \{ v_{l_k}^{(1)} \} \qquad j, k = 1, 2, \ldots, M \tag{3.68}$$

When performing this task, each node in the class layer (see Fig. 3.5) is considered to function as a "winner-takes-all" network. A kth class node receives activations only from the hidden nodes corresponding to the cases in C_k. That is, the activation received by the kth class node from the l_kth hidden node is

$$u_{kl_k}^{(2)} = v_{l_k}^{(1)} w_{kl_k}^{(1)} \tag{3.69}$$

The output of the kth class node is

$$v_{l_k}^{(2)} = \max_l \{ u_{kl_k}^{(2)} \} \tag{3.70}$$

where $v_{l_k}^{(2)}$ represents the degree of belongingness of x to class C_k. Therefore, decide that $x \in C_j$ if

$$v_j^{(2)} > v_k^{(2)} \qquad j, k = 1, 2, \ldots, M, \quad j \neq k$$

3.7.2.4 *Experimental Results*   In this section the effectiveness of the network (methodology) for automatic selection of cases is demonstrated by making the cases function as prototypes for a 1-NN classifier. Some results of investigation [71] are described when real-life vowel [35] data and medical data [73] were considered as the inputs. In all these examples, the data set was divided into two

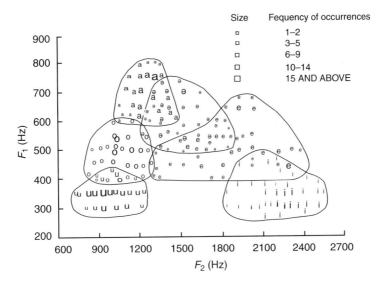

Figure 3.6   Scatter plot of the vowel data in the F_1–F_2 plane.

subsets for training and testing. The variable "perc %" denotes the proportion of the samples considered during training, and the remaining "$(100 - \text{perc})$ %" portion used for testing.

The vowel data [35] consist of a set of 871 Indian Telugu vowel sounds. These sounds were uttered in a "consonant–vowel–consonant" context by three male speakers in the age group 30 to 35 years. The data set has three features, F_1, F_2, and F_3, corresponding to the first, second, and third vowel format frequencies obtained through spectrum analysis of speech sounds. Figure 3.6 shows the overlapping nature of the six vowel classes (i.e., δ, a, i, u, e, o) in the F_1–F_2 plane (for ease of depiction). The details of the data and their extraction procedure are available in [35]. These vowel data have been used extensively for more than two decades in the area of pattern recognition research.

The medical data consist of nine input features and four pattern classes and deal with various hepatobiliary disorders [73] of 536 patient cases. The input features are the results of different biochemical tests. These tests are glutamic oxalacetic transaminate (GOT, Karmen unit), glutamic pyruvic transaminase (GPT, Karmen unit), lactate dehydrase (LDH, IU/L), gamma glutamyl transpeptidase (GGT, mU/mL), blood urea nitrogen (BUN, mg/dL), mean corpuscular volume of red blood cell (MCV, fL), mean corpuscular hemoglobin (MCH, pg), total bilirubin (TBil, mg/dL), and creatinine (CRTNN, mg/dL). The hepatobiliary disorders alcoholic liver damage (ALD), primary hepatoma (PH), liver cirrhosis (LC), and cholelithiasis (C) constitute the four classes.

Tables 3.3 and 3.4 depict some of the results obtained with the foregoing data sets for different values of λ when perc = 30 is used. The first column of these tables indicates the number of iteration(s) required by the network until it stabilizes

TABLE 3.3 **Classification Performance for Different** λ **Using Vowel Data for Perc = 30**

Number of Iterations	λ	Class	Number of Hidden Nodes	Recognition Score (%)	
				Training Set	Test Set
3	100.0	δ	21	95.24	41.18
		a	22	100.0	84.13
		i	42	98.04	72.73
		u	31	97.78	81.13
		e	53	98.39	67.59
		o	38	94.44	89.69
		Overall	207	97.30	75.0
3	150.0	δ	18	95.24	64.71
		a	13	96.15	93.65
		i	23	96.08	86.78
		u	20	88.89	96.79
		e	37	96.77	80.00
		o	26	92.59	85.71
		Overall	137	94.21	83.82
3	200.0	δ	16	80.95	64.71
		a	13	92.31	90.49
		i	21	98.04	87.60
		u	19	91.11	85.85
		e	36	93.55	81.38
		o	25	90.74	86.51
		Overall	130	92.28	83.99
1	250.0	δ	12	71.43	58.82
		a	9	88.46	80.95
		i	11	92.16	85.95
		u	9	84.44	72.64
		e	20	91.94	80.69
		o	14	81.48	74.60
		Overall	75	86.49	77.29
1	300.0	δ	10	57.14	52.94
		a	8	92.31	80.95
		i	10	92.16	86.78
		u	8	97.78	83.96
		e	20	88.71	80.69
		o	11	64.81	59.52
		Overall	67	83.78	75.82
3	350.0	δ	8	52.38	52.94
		a	7	92.31	95.24
		i	9	94.12	90.08
		u	8	97.78	89.62
		e	13	70.97	66.21
		o	8	46.30	42.86
		Overall	53	75.68	72.06

TABLE 3.3   (*Continued*)

Number of Iterations	λ	Class	Number of Hidden Nodes	Recognition Score (%) Training Set	Test Set
1	400.0	δ	8	57.14	56.86
		a	7	96.15	95.24
		i	7	88.24	86.78
		u	6	97.78	84.91
		e	10	69.35	65.52
		o	8	72.22	64.29
		Overall	46	80.31	75.16
1	450.0	δ	7	71.43	70.59
		a	5	84.62	68.25
		i	5	58.82	61.16
		u	6	93.33	83.02
		e	9	83.87	76.55
		o	6	68.52	67.46
		Overall	38	76.45	71.41

during training. It is found from these tables that the recognition scores on the training set, as expected, are higher than those on the test set. The recognition score during training decreases with an increase in the value of λ. On the other hand, for the test data, the recognition score increases with λ up to a certain value, beyond which it decreases. This can be explained as follows.

During training, the recognition score increases when there is a decrease in λ, due to a better abstraction capability although for the test data, as λ decreases, the modeling of class structures improves because there is an increase in the number of cases, and therefore the recognition score increases up to a certain value of λ. Beyond that value, as mentioned in Section 3.7.1, the number of cases with poor generalization capability (i.e., similarity functions with very small bandwidth) increases. As a result, the recognition score decreases due to overlearning.

As mentioned in Section 3.7.2.2, the number of hidden nodes in the network decreases with an increase in λ for all the example data (see Tables 3.3 and 3.4). Since class "e" of vowel data is most sparse, it needs a maximum number of cases (and hence a maximum number of hidden nodes) for its representation. This is reflected in Table 3.3. Similar observations hold good for the medical data, where the class "PH," being the sparsest, has the maximum number of hidden nodes. From these tables it can be noted that for both the data sets, the stability of the architecture of their respective networks is achieved within a very few iterations.

The effect of the size of a training set on the performance of the network is demonstrated in Table 3.5 only for the vowel data, as an example. The different values of perc considered are 10, 20, 30, 40, 50, 60, and 70, with λ = 150.0, 200.0, and 250.0. (Note that the network achieves the best generalization capability

TABLE 3.4   Classification Performance for Different λ Using the Medical Data for Perc = 30

Number of Iterations	λ	Class	Number of Hidden Nodes	Recognition Score (%) Training Set	Recognition Score (%) Test Set
1	150.0	ALD	17	61.76	32.93
		PH	30	81.13	48.80
		LC	19	91.89	73.56
		C	13	42.86	27.71
		Overall	79	71.07	46.42
1	160.0	ALD	20	71.43	56.79
		PH	35	70.37	29.84
		LC	17	78.95	66.28
		C	8	58.33	57.32
		Overall	80	69.94	50.13
1	170.0	ALD	20	71.43	58.02
		PH	34	68.52	29.03
		LC	17	78.95	66.28
		C	8	55.56	54.88
		Overall	79	68.71	49.60
1	180.0	ALD	20	68.57	56.79
		PH	34	72.22	31.45
		LC	16	71.05	61.63
		C	8	55.56	54.88
		Overall	78	67.48	49.06
1	190.0	ALD	19	80.00	61.73
		PH	33	77.78	36.29
		LC	14	76.32	68.60
		C	6	8.33	12.20
		Overall	72	62.58	43.97
7	200.0	ALD	15	76.47	58.54
		PH	25	71.70	45.60
		LC	14	72.97	68.97
		C	11	25.71	9.64
		Overall	137	62.89	45.89

TABLE 3.5   Classification Performance for Different λ and Perc on Vowel Data

λ	Recognition Score (%) Perc = 10	Perc = 20	Perc = 30	Perc = 40	Perc = 50	Perc = 60	Perc = 70
150.0	70.99	81.55	83.82	80.42	84.67	86.29	87.01
200.0	75.70	82.12	83.99	83.27	84.67	85.71	86.20
250.0	75.06	80.11	77.29	80.23	82.38	83.43	84.03

for $\lambda = 200.0$; see Table 3.3.) Table 3.5 shows that the recognition score on the test set generally increases with the size of the training set, as expected.

The performance of the classifier (where the cases are considered as prototypes) was also compared with that of the following ones:

- A standard k-NN classifier with $k = \sqrt{m}$ (m being the number of training samples), where all the perc % samples, selected randomly, are considered as prototypes. (It is known that as m goes to infinity, if the values of k and k/m can be made to approach infinity and zero, respectively, the performance of the k-NN classifier approaches that of the (optimal) Bayes classifier [25]. One such value of k for which these limiting conditions are satisfied is $\sqrt{m}$.)
- Bayes' maximum likelihood classifier, where a multivariate normal distribution of samples with different class dispersion matrices and a priori probabilities ($= m_j/m$, for m_j patterns from class C_j) are assumed, and all of the perc % samples are used to compute the mean vectors and the covariance matrices.

Table 3.6 depicts that the network (CBNN) performs better than the k-NN ($k = \sqrt{m}$) and Bayes' maximum likelihood classifiers for vowel data. In the case of medical data, while the performance of CBNN on the training set is better than those obtained by the others, the reverse is true for the test samples.

TABLE 3.6   Recognition Score of Various Classifiers on Different Data Sets

| | | Recognition Score (%) | | | | | |
| | | CBNN | | Bayes | | k-NN | |
Data Set	Class	Training	Testing	Training	Testing	Training	Testing
Pat1	1	100.0	100.0	100.0	100.0	100.0	99.38
	2	100.0	100.0	34.48	19.12	96.55	100.0
	Overall	100.0	100.0	88.62	85.90	99.40	99.49
Vowel	δ	80.95	64.71	38.10	43.14	23.81	33.33
	a	92.31	90.48	88.46	85.71	80.77	85.71
	i	98.04	87.60	90.20	85.12	88.24	85.12
	u	91.11	95.85	91.11	90.57	86.67	76.42
	e	93.55	81.38	75.81	90.69	75.81	77.93
	o	90.74	96.51	92.59	85.71	92.59	88.89
	Overall	92.28	83.99	83.01	81.70	79.92	78.43
Medical	ALD	71.43	56.79	61.76	50.00	52.94	46.34
	PH	70.37	29.84	54.72	64.80	69.81	77.60
	LC	78.95	66.28	51.35	36.78	21.62	29.89
	C	58.33	57.32	91.43	75.90	54.29	61.45
	Overall	69.94	50.13	63.52	57.56	51.57	56.23

3.8   CASE SELECTION USING ROUGH SELF-ORGANIZING MAP

In Section 3.7 we explained how a neuro-fuzzy framework can be formulated for the selection of cases (prototypes) under the supervised mode. The present section deals with the methodology of Pal et al. [64], demonstrating an integration of rough sets, fuzzy sets, and self-organizing networks for extracting class prototypes (representative cases for the entire data set) under the unsupervised mode of learning. One may note here that a similar integration of these three soft computing tools under supervised mode was mentioned in Section 2.3.3, where it is shown how different class regions can be represented with fuzzy membership functions of a varying number for generating representative cases in terms of only informative regions and relevant features (i.e., reduced subsets of original attributes)—thereby enabling fast case retrieval.

Rough set theory [74] provides an effective means for classificatory analysis of data tables. The main goal of rough set–theoretic analysis is to synthesize or construct approximations (upper and lower) of concepts from the acquired data. The key concepts here are those of information granule and reducts. *Information granule* formalizes the concept of finite-precision representation of objects in real-life situations, and the *reducts* represent the core of an information system (in terms of both objects and features) in a granular universe. An important use of rough set theory has been in generating logical rules for classification and association [75]. These logical rules correspond to different important granulated regions of the feature space, which represent data clusters. For application of rough sets in pattern recognition and data mining problems, one may refer to the recent special issue [76].

A self-organizing map (SOM) [77] is an unsupervised network that has recently become popular for unsupervised mining of large data sets. The process of self-organization generates a network whose weights represent prototypes of the input data. These prototypes may be considered as cases representing the entire data set. Unlike those produced by existing case generation methodologies, they are not just a subset of the original data but evolved in the self-organizing process. Since SOM suffers from the problem of slow convergence and local minima, a synergistic integration of rough set theory with SOM offers a fast and robust solution to the initialization and local minima problem, thereby designing rough SOM (RSOM). Here rough set theory is used to encode the domain knowledge in the form of crude rules, which are mapped for initialization of weights as well as for determination of the network size. Fuzzy set theory is used for discretization of feature space. Performance of the network is measured in terms of learning time, representation error, clustering quality, and network compactness. All these characteristics have been demonstrated experimentally and compared with that of the conventional SOM.

3.8.1   Pattern Indiscernibility and Fuzzy Discretization of Feature Space

A primary notion of rough set is of indiscernibility relation. For continuous-valued attributes, the feature space needs to be discretized for defining indiscernibility relations and equivalence classes. Discretization is a widely studied problem in rough set theory, and fuzzy set theory was used for effective discretization. Use of fuzzy sets has several advantages over "hard" discretization, such as modeling of overlapped clusters and linguistic representation of data [78]. Here each feature is discretized into three levels: low, medium, and high; finer discretizations may lead to better accuracy at the cost of higher computational load.

As mentioned in Sections 2.3.2 and 3.4.2, each feature of a pattern is described in terms of their fuzzy membership values in the linguistic property sets "low" (L), "medium" (M), and "high" (H). Let these be represented by L_j, M_j, and H_j, respectively. The features for the ith pattern F_i are mapped to the corresponding three-dimensional feature space of $\mu_{\text{low}(F_{i,j})}(\hat{e}_i)$, $\mu_{\text{medium}(F_{i,j})}(\hat{e}_i)$, and $\mu_{\text{high}(F_{i,j})}(\hat{e}_i)$ by equation (3.33). An n-dimensional pattern $\hat{e}_i = [F_{i1}, F_{i2}, \ldots, F_{in}]$ is thus represented as a $3n$-dimensional vector [42].

This effectively discretizes each feature into three levels. Then consider only those attributes that have a numerical value greater than some threshold Th ($= 0.5$, say). This implies clamping only those features demonstrating high membership values with unity while the others are fixed at zero. An attribute-value table is constructed comprising the binary-valued $3n$-dimensional feature vectors above. Here we mention that the representation of linguistic fuzzy sets by Π-functions and the procedure in selecting the centers and radii of the overlapping fuzzy sets are the same as in Section 2.3.3.1. The nature of the functions is also the same as in Figure 2.13.

3.8.2 Methodology for Generation of Reducts

After the binary membership values are obtained for all the patterns, the decision table has been constituted for rough set rule generation. Let there be m sets $O_1, O_2, \ldots, O_m$ of objects in the attribute-value table (obtained by the procedure described in Section 3.8.1) having identical attribute values, and $\text{card}(O_i) = n_{ki}$, $i = 1, 2, \ldots, m$ such that $n_{k1} \geq n_{k2} \geq \cdots \geq n_{km}$ and $\sum_{i=1}^{m} n_{ki} = n_k$. As in Section 2.3.3.3, the attribute–value table is represented as an $m \times 3n$ array. Let $n_{k'_1}, n_{k'_2}, \ldots, n_{k'_m}$ denote the distinct elements among $n_{k1}, n_{k2}, \ldots, n_{km}$ such that $n_{k'_1} > n_{k'_2} > \cdots > n_{k'_m}$. Let a heuristic threshold Tr be defined as in equation (2.14), so that all entries having frequency less than Tr are eliminated from the table, resulting in the reduced attribute–value table $\hat{S}$. Note that the main motive of introducing this threshold function lies in reducing the size of the model. One attempts to eliminate noisy pattern representatives (having lower values of n_{ki}) from the reduced attribute–value table. From the reduced attribute–value table obtained, reducts are determined using the methodology described below.

Let $\{x_{i1}, x_{i2}, \ldots, x_{ip}\}$ be the set of those objects of U that occur in $\hat{S}$. Now a discernibility matrix [denoted $M(B)$] is defined as follows [64,79]:

$$c_{ij} = \{\text{Attr} \in B \, : \, \text{Attr}(x_i) \neq \text{Attr}(x_j)\} \qquad \text{for } i, j = 1, 2, \ldots, n \qquad (3.71)$$

For each object $x_j \in \{x_{i1}, x_{i2}, \ldots, x_{ip}\}$, the discernibility function f_{x_j} is defined as

$$f_{x_j} = \wedge\{\vee(c_{ij}) \ : \ 1 \le i,j \le n, j < i, c_{ij} \ne \phi\} \tag{3.72}$$

where $\vee(c_{ij})$ is the disjunction of all members of c_{ij}. One thus obtains a rule r_i, that is, $g_i \rightarrow cluster_i$, where g_i is the disjunctive normal form of $f_{x_j}, j \in i_1, i_2, \ldots, i_p$.

3.8.3   Rough SOM

3.8.3.1   *Self-Organizing Maps*   A Kohonen feature map is a two-layered network. The first layer of the network is the input layer. The second layer, called the *competitive layer*, is usually organized as a two-dimensional grid. All interconnections go from the first layer to the second as in Figure 3.7. For details, see Appendix B. All the nodes in the competitive layer compare the inputs with their weights and compete with each other to become the winning unit having the lowest difference. The basic idea underlying *competitive learning* is roughly as follows: Assume a sequence of input vectors $\{\hat{e} = \hat{e}(t) \in \mathbf{R}_n$, where t is the time coordinate$\}$ and a set of variable reference vectors $\{w_i(t) \ : \ w_i \in \mathbf{R}_n, i = 1, 2, \ldots, k$, where k is the number of units in the competitive layer$\}$. Initially, the values of the reference vectors (also called *weight vectors*) are set randomly. At each successive instant of time t, an input pattern $\hat{e}(t)$ is presented to the network. The input pattern $\hat{e}(t)$ is then compared with each $w_i(t)$ and the best-matching $w_i(t)$ is updated to match the current $\hat{e}(t)$ even more closely. If the comparison is based on some distance measure $d(\hat{e}, w_i)$, altering w_i must be such that if $i = c$, the index of the best-matching reference vector, then $d(\hat{e}, w_c)$ is reduced and all the other reference vectors w_i, with $i \ne c$, are left intact. In this way the various reference vectors tend to become specifically "tuned" to different domains of the input variable $\hat{e}$.

The first step in the operation of a Kohonen network is to compute a matching value for each unit in the competitive layer. This value measures the extent to which

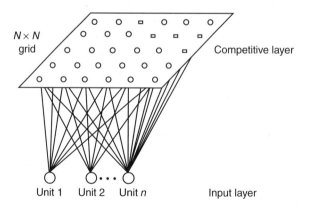

Figure 3.7   Basic network structure for a Kohonen feature map.

the weights or reference vectors of each unit match the corresponding values of the input pattern. The matching value for each unit i is $\|\hat{e} - w_i\|$, which is the distance between vectors $\hat{e}$ and w_i and is computed by

$$\sqrt{\sum_j (\hat{e}_j - w_{ij})^2} \qquad \text{for} \quad j = 1, 2, \ldots, n \qquad (3.73)$$

The unit with the lowest matching value (the best match) wins the competition. In other words, the unit c is said to be the best-matched unit if

$$\|\hat{e} - w_c\| = \min_i\{\|\hat{e} - w_i\|\}, \qquad i = 1, 2, \ldots, (N \times N) \qquad (3.74)$$

where the minimum is taken over all the $N \times N$ units in the competitive layer. If two units have the same matching value, then by convention, the unit with the lower index value i is chosen.

The next step is to self-organize a two-dimensional map that reflects the distribution of input patterns. In biophysically inspired neural network models, correlated learning by spatially neighboring cells can be implemented using various kinds of lateral feedback connections and other lateral interactions. Here the lateral interaction is enforced directly in a general form, for arbitrary underlying network structures, by defining a neighborhood set N_c around the winning cell. At each learning step, all the cells within N_c are updated, whereas cells outside N_c are left intact. The update equation is

$$\Delta w_{ij} = \begin{cases} \lambda(\hat{e}_j - w_{ij}) & \text{if unit } i \text{ is in the neighborhood } N_c \\ 0 & \text{otherwise} \end{cases} \qquad (3.75)$$

and

$$w_{ij}^{\text{new}} = w_{ij}^{\text{old}} + \Delta w_{ij} \qquad (3.76)$$

Here λ is the learning parameter. This adjustment results in both the winning unit and its neighbors, having their weights modified, becoming more like the input pattern. The winner then becomes more likely to win the competition should the same or a similar input pattern be presented subsequently.

3.8.3.2 *Incorporation of Rough Sets in SOM*   As described in Section 3.8.2, the dependency rules generated using rough set theory from an information system are used to discern objects with respect to their attributes. However, the dependency rules generated by rough set are coarse and therefore need to be fine-tuned. Here the dependency rules are used to get a crude knowledge of the cluster boundaries of the input patterns to be fed to a self-organizing map. This crude knowledge is used to encode the initial weights of the nodes of the map, which is then trained using the usual learning process (Section 3.8.3.1). Since an

initial knowledge of the cluster boundaries is encoded into the network, the learning time is reduced greatly, with improved performance.

The steps involved in the process, as formulated by Pal et al. [64], are summarized below.

Step 1. From the initial data set, use the fuzzy discretization process to create the information system.

Step 2. For each object in the information table, generate the discernibility function

$$f_A(a_1, a_2, \ldots, a_{3n}) = \wedge\{\vee c_{ij} \; : \; 1 \leq i,j \leq n, j < i, c_{ij} \neq \emptyset\} \qquad (3.77)$$

where $a_1, a_2, \ldots, a_{3n}$ are the $3n$ Boolean variables corresponding to the attributes $\text{Attr}_1, \text{Attr}_2, \ldots, \text{Attr}_{3n}$ of each object in the information system. The expression f_A is reduced to the set of all prime implicants of f_A that determines the set of all reducts of A.

Step 3. The self-organizing map is created with $3n$ inputs (Section 3.8.1), which correspond to the attributes of the information table, and a competitive layer of $N \times N$ grid of units where N is the total number of implicants present in discernibility functions of all the objects of the information table.

Step 4. Each implicant of the function f_A is mapped to a unit in the competitive layer of the network, and high weights are given to those links that come from the attributes, which occur in the implicant expression. The idea behind this is that when an input pattern belonging to an object, say O_i, is applied to the inputs of the network, one of the implicants of the discernibility function of O_i will be satisfied, and the corresponding unit in the competitive layer will fire and emerge as the winning unit. All the implicants of an object O_i are placed in the same layer, while the implicants of different objects are placed in different layers separated by the maximum neighborhood distance. In this way the initial knowledge obtained with rough set methodology is used to train the SOM.

This is explained with the following example. Let the reduct of an object O_i be

$$O_i \; : \; (F_{1\text{low}} \wedge F_{2\text{medium}}) \vee (F_{1\text{high}} \wedge F_{2\text{high}})$$

where $F_{(\cdot)\text{low}}, F_{(\cdot)\text{medium}},$ *and* $F_{(\cdot)\text{high}}$ represent the low, medium, and high values of the corresponding features. Then the implicants are mapped to the nodes of the layer as shown in Figure 3.8. Here high weights (H) are given only to those links that come from the features present in the implicant expression. Other links are given low weights.

3.8.4   Experimental Results

Let us explain here some of the results of investigation [64] demonstrating the effectiveness of the methodology on an artificially generated data set (Fig. 3.9)

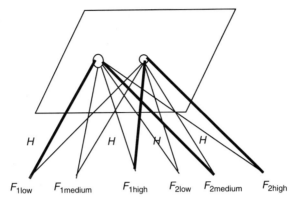

Figure 3.8 Mapping of reducts in the competitive layer of RSOM.

and two sets of real-life data, the speech data "vowel" and the medical data (which are described in Section 3.7.2.4). The artificial data of Figure 3.9 consist of two features containing 417 points from two horseshoe-shaped clusters. Vowel data has three features and 871 samples from six classes, while the medical data have nine features and deals with hepatobiliary disorders of 536 patients from four classes.

The following quantities were considered for comparing the performance of the RSOM with that of the randomly initialized self-organized map.

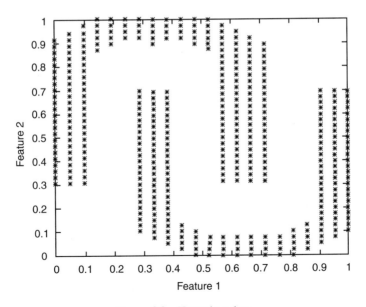

Figure 3.9 Horseshoe data.

1. *Quantization error.* The quantization error (q_E) measures how fast the weight vectors of the winning nodes in the competitive layer are aligning themselves with the input vectors presented during training. It is calculated using the equation

$$q_E = \frac{\sum_{p=1}^{n} \left[\sum_{\text{all wining nodes}} \sqrt{\sum_j (x_{pj} - w_j)^2} \right]}{\text{number of patterns}} \qquad (3.78)$$

Here $j = 1, 2, \ldots, m$ (m being the number of input features to the net), x_{pj} is the jth component of the pth pattern, and n is the total number of patterns. Hence, the higher the quantization error (q_E), the greater the difference between the reference vectors and the input vectors of the nodes in the competitive layer.

2. *Entropy and β-index.* The entropy measure [23] and β-index [80] reflect the quality of the cluster structure.

Entropy: Let the distance between two weight vectors p and q be

$$d(p, q) = \left[\sum_j \left(\frac{x_{pj} - x_{qj}}{\text{max}_j - \text{min}_j} \right)^2 \right]^{1/2} \qquad (3.79)$$

where x_{pj} and x_{qj} denote the weight values for p and q, respectively, along the jth direction, and $j = 1, 2, \ldots, m$, m being the number of features input to the net; max_j and min_j are, respectively, the maximum and minimum values computed over all the samples along the jth axis. Let the similarity between p and q be defined as

$$\text{SM}(p, q) = e^{-\beta d_{pq}} \qquad (3.80)$$

where $\beta = -\ln 0.5 / \bar{d}$, a positive constant such that

$$\text{SM}(p, q) = \begin{cases} 1 & \text{if } d(p, q) = 0 \\ 0 & \text{if } d(p, q) = \infty \\ 0.5 & \text{if } d(p, q) = \bar{d} \end{cases}$$

where $\bar{d}$ is the average distance between points computed over the entire data set. *Entropy* is defined as

$$E = -\sum_{p=1}^{l} \sum_{q=1}^{l} (\text{SM}(p, q) \log \text{SM}(p, q) + (1 - \text{SM}(p, q)) \log(1 - \text{SM}(p, q))) \qquad (3.81)$$

If the data are uniformly distributed in the feature space, entropy is maximum. When the data have well-formed clusters, the uncertainty is low and so is the entropy.

β-index: The *β-index* [80] is defined as

$$\beta = \frac{\sum_{i=1}^{k} \sum_{p=1}^{n_i} (\hat{e}_p^i - \bar{\bar{e}})^T (\hat{e}_p^i - \bar{\bar{e}})}{\sum_{i=1}^{k} \sum_{p=1}^{n_i} (\hat{e}_p^i - \bar{e}^i)^T (\hat{e}_p^i - \bar{e}^i)} \tag{3.82}$$

where n_i is the number of points in the *i*th $(i = 1, 2, \ldots, k)$ cluster; $\hat{e}_p^i$ the *p*th pattern $(p = 1, 2, \ldots, n_i)$ in cluster *i*; $\bar{e}^i$ the mean of n_i patterns of the *i*th cluster, $\sum_i n_i = n$, where *n* is the total number of patterns; and $\bar{\bar{e}}$ is the mean value of the entire set of patterns. Note that β is nothing but the ratio of the total variation and within-cluster variation. This type of measure is widely used for feature selection and cluster analysis. For a given data and *k* (number of clusters) value, the greater the homogeneity within the clustered regions, the higher the β value would be.

3. *Frequency of winning nodes* (f_k). This frequency is defined as the number of winning top *k* nodes (f_k) in the competitive layer, where *k* is the number of rules (characterizing the clusters) obtained using rough sets. Here $k = 4$ for horseshoe data, $k = 14$ for vowel data, and $k = 7$ for medical data. f_k reflects the error if all but *k* nodes would have been pruned. In other words, it measures the number of sample points represented correctly by these nodes.

4. *Number of iterations.* This indicates the number of iterations at which the error does not change much.

The comparative results for the three data sets are presented in Table 3.7. The following conclusions [64] can be made from the results obtained:

1. *Better cluster quality.* As seen from Table 3.7, RSOM has a lower entropy value, thus implying a lower intracluster distance and a higher intercluster distance in the clustered space compared to conventional SOM. RSOM also has a higher β-index value, indicating more homogeneity within its clustered regions. The quantization error of RSOM is also far less than that of SOM.

2. *Less learning time.* The number of iterations required to achieve the error level is far less in RSOM than in SOM. The convergence curves of the quantization

TABLE 3.7   Comparison of RSOM with SOM

Data	Initialization	Quantization Error	Iteration at Which Error Converged	Entropy	f_k	β-Index
Horseshoe	Random	0.038	5000	0.7557	83	0.99
	Rough	0.022	50	0.6255	112	0.99
Vowel	Random	32.588	8830	0.6717	245	0.06
	Rough	0.081	95	0.6141	316	0.96
Medical	Random	28.855	8666	0.6744	110	0.61
	Rough	0.246	102	0.6121	125	0.71

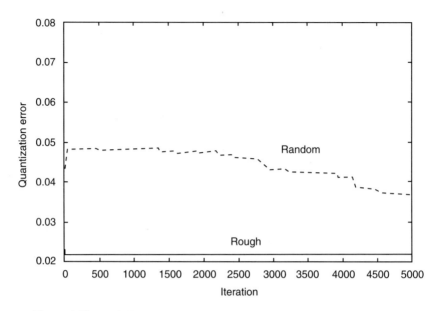

Figure 3.10   Variation of quantization error with iteration for horseshoe data.

errors are presented in Figures 3.10 to 3.12 for the data sets used. It is seen that RSOM starts from a very low value of quantization error compared to SOM.

3. *Compact representation of data.* It is seen for RSOM, fewer nodes in the competitive layer dominate (i.e., they win for most of the samples in the training set). On the other hand, in conventional SOM this number is higher. This is

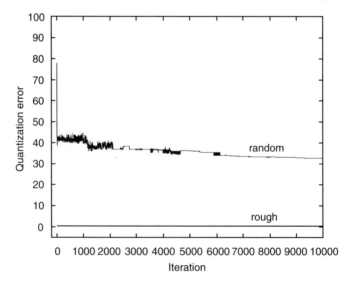

Figure 3.11   Variation of quantization error with iteration for vowel data.

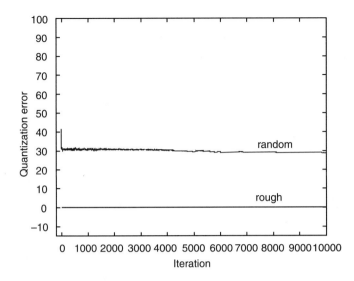

Figure 3.12 Variation of quantization error with iteration for medical data.

quantified by the frequency of winning of the top k nodes. It is observed that this value is much higher for RSOM, thus signifying less error if all but k nodes would have been pruned. In other words, RSOM achieves a more compact representation of the data. Therefore, as represented by the weight vectors of the winning nodes,

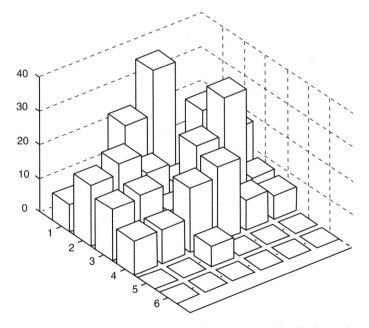

Figure 3.13 Frequency of winning nodes using random weights for horseshoe data.

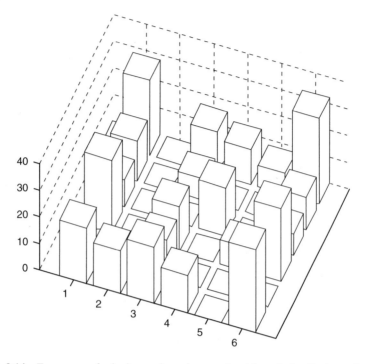

Figure 3.14   Frequency of winning nodes using rough set knowledge for horseshoe data.

the cases constitute a compact case base. Since RSOM achieves compact clusters, this will enable one to extract nonambiguous rules. As a demonstration of the nature of distribution of the frequency of winning nodes, the results for the horseshoe data, corresponding to SOM and RSOM are shown in Figures 3.13 and 3.14. Separation between the clusters is seen to be more prominent in Figure 3.14. These winning nodes may be viewed as the prototype points (cases) representing the two classes. Unlike the conventional methods, the cases and prototypes selected here are not just a subset of the original data points, but represent some collective information generated by the network after learning the entire data set.

3.9   SUMMARY

In this chapter we have explained various similarity measures and the notion of fuzzy set theory in measuring them. This was followed by different algorithms and methodologies for classification/clustering of cases, determination of feature weights, case matching, case selection, and case retrieval. Both supervised and unsupervised modes of learning have been considered wherever possible. Various soft computing approaches using fuzzy logic, genetic algorithms, artificial neural networks and rough sets, and hybrid tools such as fuzzy-neural and rough-fuzzy-neural have been described together with their relevance. The merits of using integrations in a soft computing paradigm have been illustrated with experimental results.

REFERENCES

1. B. Smyth and M. T. Keane, Adaptation-guided retrieval: questioning the similarity assumption in reasoning, *Artificial Intelligence*, vol. 102, pp. 249–293, 1998.

2. M. T. Keane, *Analogical Problem Solving*, Ellis Horwood, Chichester, West Sussex, England, 1988.

3. M. T. Keane, Analogical asides on case-based reasoning, in *Topics in Case-Based Reasoning* (S. Wess, K. D. Althoff, and M. M. Richter, eds.), Springer-Verlag, Berlin, pp. 21–32, 1994.

4. D. L. Medin, R. L. Goldstone, and D. Gentner, Respects for similarity, *Psychological Review*, vol. 100, no. 2, pp. 254–278, 1993.

5. G. L. Murphy and D. L. Medin, The role of theories in conceptual coherence, *Psychology Review*, vol. 92, pp. 289–316, 1995.

6. W. Cheetham, Case-based reasoning with confidence, in *Proceedings of the Fifth European Workshop on Case-Based Reasoning* (*EWCBR-00*), Trento, Italy, Springer-Verlag, Berlin, pp. 15–25, 2000.

7. I. Watson and S. Perera, The evaluation of a hierarchical case representation using context guided retrieval, in *Proceedings of the Second International Conference on Case-Based Reasoning* (*ICCBR-97*), Providence, RI, Springer-Verlag, Berlin, pp. 255–266, 1997.

8. G. Falkman, Similarity measures for structured representations: a definitional approach, in *Proceedings of the Fifth European Workshop on Case-Based Reasoning* (*EWCBR-00*), Trento, Italy, Springer-Verlag, Berlin, pp. 380–392, 2000.

9. T. W. Liao, Z. Zhang, and C. R. Mount, Similarity measures for retrieval in case-based reasoning systems, *Applied Artificial Intelligence*, vol. 12, pp. 267–288, 1998.

10. R. W. Hamming, Error detecting and error correcting codes, *Bell System Technical Journal*, vol. 9, pp. 147–160, 1950.

11. R. W. Soukoreff and I. S. MacKenzie, Measuring errors in text entry tasks: an application of the Levenshtein string distance statistic, *Extended Abstracts of the ACM Conference on Human Factors in Computing System* (*CHI 2001*), ACM, New York, pp. 319–320, 2001.

12. M. R. Anderberg, *Cluster Analysis for Applications*, Academic Press, San Diego, CA, 1973.

13. A. Tversky, Features of similarity, *Psychological Review*, vol. 84, no. 4, pp. 327–352, 1977.

14. M. Sebag and M. Schoenauer, A rule-based similarity measure, in *Proceedings of the First European Workshop* (*GWCBR-93*), Kaiserslautern, Germany, Springer-Verlag, Berlin, pp. 119–131, 1993.

15. G. Weber, Examples and reminders in a case-based help system, in *Proceedings of the Second European Workshop* (*EWCBR-94*), Chantilly, France, Springer-Verlag, Berlin, pp. 165–177, 1994.

16. J. E. Hunt, D. E. Cooke, and H. Holstein, Case memory and retrieval based on the immune system, in *Proceedings of the First International Conference on Case-Based Reasoning* (*ICCBR-95*), Sesimbra, Portugal, Springer-Verlag, Berlin, pp. 205–216, 1995.

17. S. K. Pal, Fuzzy image processing and recognition: uncertainties handling and applications (invited paper), *International Journal of Image Graphics*, vol. 1, no. 2, pp. 165–195, 2001.

18. G. Agre, KBS maintenance as learning two-tiered domain representation, in *Proceedings of the First International Conference on Case-Based Reasoning (ICCBR-95)*, Sesimbra, Portugal, Springer-Verlag, Berlin, pp.109–120, 1995.

19. T. P. Cain, M. J. Pazzani, and G. Silverstein, Using domain knowledge to influence similarity judgments, in *Proceedings of the 1991 DARPA Workshop on Case-Based Reasoning*, San Francisco, Morgan Kaufmann, San Francisco, pp. 191–199, 1991.

20. E. Armengol and E. Plaza, Similarity for relational CBR, in *Proceedings of the Fourth International Conference on Case-Based Reasoning (ICCBR-01)*, Vancouver, British Columbia, Canada, Springer-Verlag, Berlin, pp. 44–58, 2001.

21. E. K. Burke, B. MacCarthy, S. Petrovic, and R. Qu, Case-based reasoning in course timetabling: an attribute graph approach, in *Proceedings of the Fourth International Conference on Case-Based Reasoning (ICCBR-01)*, Vancouver, British Columbia, Canada, Springer-Verlag, Berlin, pp. 90–104, 2001.

22. R. Bergmann and A. Stahl, Similarity measures for object-oriented case representation, in *Proceedings of the Fourth European Workshop on Case-Based Reasoning (EWCBR-98)*, Dublin, Ireland, Springer-Verlag, Berlin, pp. 25–36, 1998.

23. S. K. Pal and S. Mitra, *Neuro-Fuzzy Pattern Recognition: Methods in Soft Computing*, Wiley, New York, 1999.

24. S. K. Pal and A. Pal (eds.), *Pattern Recognition: From Classical to Modern Approaches*, World Scientific, Singapore, 2001.

25. K. Fukunaga, *Introduction to Statistical Pattern Recognition*, Academic Press, San Diego, CA, 1972.

26. L. A. Zadeh, Fuzzy sets, *Information Control*, vol. 8, pp. 338–353, 1965.

27. L. A. Zadeh, From search engines to question-answering systems: the need for new tools, in *Proceedings of the Twelfth IEEE International Conference on Fuzzy Systems (FUZZ-IEEE-03)*, St. Louis, MO, Morgan Kaufmann, San Francisco, pp. 1107–1109, 2003.

28. J. M. Mendel, Fuzzy sets for words: a new beginning, in *Proceedings of the Twelfth IEEE International Conference on Fuzzy Systems (FUZZ-IEEE-03)*, St. Louis, MO, Morgan Kaufmann, San Francisco, pp. 37–42, 2003.

29. T. R. Davies and S. J. Russell, A logical approach to reasoning by analogy, in *Proceedings of the Tenth International Joint Conference on Artificial Intelligence (IJCAI-87)*, Milan, Italy, Morgan Kaufmann, San Francisco, pp. 264–270, 1987.

30. I. Gilboa and D. Schmeidler, Case-based decision theory, *Quarterly Journal of Economics*, vol. 110, pp. 605–639, 1995.

31. D. Dubois, F. Esteva, P. Garcia, L. Godo, R. Lopez de Mantaras, and H. Prade, Fuzzy modeling of case-based reasoning and decision, in *Proceedings of the Second International Conference on Case-Based Reasoning (ICCBR-97)*, Providence, RI, Springer-Verlag, Berlin, pp. 599–610, 1997.

32. M. Q. Xu, K. Hirota, and H. Yoshino, A fuzzy theoretical approach to case-based representation and inference in CISG, *Artificial Intelligence and Law*, vol. 7, pp. 259–272, 1999.

33. M. Setnes, R. Babuska, U. Kaymak, L. Nauta, and H. R. Van, Similarity measures in fuzzy rule base simplification, *IEEE Transactions on Systems, Man, and Cybernetics, Part B: Cybernetics*, vol. 28, no. 3, pp. 376–386, 1998.

34. T. Y. Slonim and M. Schneider, Design issues in fuzzy case-based reasoning, *Fuzzy Sets and Systems*, vol. 117, pp. 251–267, 2001.

35. S. K. Pal and D. Dutta Majumder, *Fuzzy Mathematical Approach to Pattern Recognition*, Wiley (Halsted), New York, 1986.

36. S. K. Pal and A. Skowron (eds.), *Rough Fuzzy Hybridization: A New Trend in Decision Making*, Springer-Verlag, Singapore, 1999.

37. M. Banerjee, S. Mitra, and S. K. Pal, Rough fuzzy MLP: knowledge encoding and classification, *IEEE Transaction on Neural Networks*, vol. 9, pp. 1203–1216, 1998.

38. J. C. Bezdek and S. K. Pal, *Fuzzy Models for Pattern Recognition: Methods That Search for Structures in Data*, IEEE Press, Piscataway, NJ, 1992.

39. A. Eichhorn, D. Girimonte, A. Klose, and R. Kruse, Neuro-fuzzy classification of surface form deviations, in *Proceedings of the Twelfth IEEE International Conference on Fuzzy Systems (FUZZ-IEEE-03)*, St. Louis, MO, Morgan Kaufmann, San Francisco, pp. 902–907, 2003.

40. A. Soria-Frisch, Hybrid SOM and fuzzy integral frameworks for fuzzy classification, in *Proceedings of the Twelfth IEEE International Conference on Fuzzy Systems (FUZZ-IEEE-03)*, St. Louis, MO, Morgan Kaufmann, San Francisco, pp. 840–845, 2003.

41. P. K. Singal, S. Mitra, and S. K. Pal, Incorporation of fuzziness in ID3 and generation of network architecture, *Neural Computing and Applications*, vol. 10, pp. 155–164, 2001.

42. S. K. Pal and S. Mitra, Multi-layer perceptron, fuzzy sets and classification, IEEE *Transactions on Neural Networks*, vol. 3, pp. 683–697, 1992.

43. D. P. Mandal, C. A. Murthy, and S. K. Pal, Formulation of a multi-valued recognition system, *IEEE Transactions on Systems, Man, and Cybernetics*, vol. 22, pp. 607–620, 1992.

44. J. Dunn, A fuzzy relative of the isodata process and its use in detecting compact, well-separated clusters, *Journal of Cybernetics*, vol. 3, no. 3, pp. 32–57, 1973.

45. J. C. Bezdek, *Pattern Recognition with Fuzzy Objective Function Algorithms*, Plenum Press, New York, 1981.

46. Z. Chi, H. Yan, and T. Pham, *Fuzzy Algorithms: With Applications to Image Processing and Pattern Recognition*, World Scientific, Singapore, 1996.

47. J. Basak, R. K. De, and S. K. Pal, Fuzzy feature evaluation index and connectionist realization II. Theoretical analysis, *Information Sciences*, vol. 11, pp. 1–17, 1998.

48. H. Frigui and S. Salem, Fuzzy clustering and subset feature weighting, in *Proceedings of the Twelfth IEEE International Conference on Fuzzy Systems (FUZZ-IEEE-03)*, St. Louis, MO, Morgan Kaufmann, San Francisco, pp. 857–862, 2003.

49. R. K. De, J. Basak, and S. K. Pal, Unsupervised feature extraction using neuro-fuzzy approach, *Fuzzy Sets and Systems*, vol. 126, pp. 277–291, 2002.

50. W. Dubitzky and F. Azuaje, A genetic algorithm and growing cell structure approach to learning case retrieval structures, in *Soft Computing in Case Based Reasoning* (S. K. Pal, T. S. Dillon, and D. S. Yeung, eds.), Springer-Verlag, London, pp. 115–146, 2001.

51. S. K. Pal, R. K. De, and J. Basak, Unsupervised feature evaluation: a neuro-fuzzy approach, *IEEE Transactions on Neural Networks*, vol. 11, no. 2, pp. 366–376, 2000.

52. R. K. De, J. Basak, and S. K. Pal, Neuro-fuzzy feature evaluation with theoretical analysis, *Neural Networks*, vol. 12, pp. 1429–1455, 1999.

53. J. H. Holland, *Adaptation in Natural and Artificial Systems*, University of Michigan Press, Ann Arbor, MI, 1975.

54. D. E. Goldberg, *Genetic Algorithms in Search, Optimization, and Machine Learning*, Addison-Wesley, Reading, MA, 1989.

55. J. S. R. Jang, C. T. Sun, and E. Mizutani, *Neuro-Fuzzy and Soft Computing: A Computational Approach to Learning and Machine Intelligence*, Prentice Hall, Upper Saddle River, NJ, 1997.

56. P. Thrift, A neural network model for case-based reasoning, in *Proceedings of the Second DARPA Workshop on Case-Based Reasoning*, Pensacola Beach, FL, pp. 334–337, 1989.

57. B. Yao and Y. He, A hybrid system for case-based reasoning, in *Proceedings of the World Congress on Neural Networks*, San Diego, CA, pp. 442–446, 1994.

58. P. B. Musgrove, J. Davis, and D. Izzard, A comparison of nearest neighbor, rule induction and neural networks for the recommendation of treatment at anticoagulant out-patient clinics, in *Proceedings of the Second U.K. Workshop on Case-Based Reasoning* (I. Watson, ed.), Salford University, Salford, Lancashire, England, pp. 55–70, 1996.

59. E. Domeshek, A case study of case indexing: designing index feature sets to suit task demands and support parallelism, in *Advances in Connectionist and Neural Computation Theory* (J. Barnden and K. Holyoah, eds.), vol. 2, *Analogical Connections*, Ablex, Stamford, CT, pp, 126–168, 1993.

60. M. Malek, A connectionist indexing approach for CBR systems, in *Case-Based Reasoning Research and Development* (M. Veloso and A. Aamodt, eds.), Springer-Verlag, Berlin, pp. 520–527, 1995.

61. J. Main, T. S. Dillon, and R. Khosla, Use of fuzzy feature vector and neural nets for case retrieval in case based systems, *Proceedings of NAFIPS*, pp. 438–443, 1996.

62. C. Milare and A. Decarvalho, Using a neural network in a CBR system, in *Proceedings of the International Conference on Computational Intelligence and Multimedia Applications (ICCIMA-98)*, Victoria, Australia, pp. 43–48, 1998.

63. E. Reategui, J. A. Campbell, and S. Borghetti, Using a neural network to learn general knowledge in a case-based system, in *Case-Based Reasoning Research and Development* (M. Veloso and A. Aamodt, eds.), Springer-Verlag, Berlin, pp. 528–537, 1995.

64. S. K. Pal, B. Dasgupta, and P. Mitra, Rough-self organizing map, *Applied Intelligence* (still to appear).

65. P. Mitra, S. Mitra, and S. K. Pal, Evolutionary modular MLP with rough sets and ID3 algorithm for staging of cervical cancer, *Neural Computing and Applications*, vol. 10, pp. 67–76, 2001.

66. R. Beal and T. Jackson, *Neural Computing: An Introduction*, Adam Hilger, Bristol, Gloucastershire, England, 1990

67. A. De Luca and S. Termin, A definition of a nonprobabilistic entropy in the setting of fuzzy set theory, *Information and Control*, vol. 20, pp. 301–312, 1972.

68. M. Higashi and G. J. Klir, Measures on uncertainty and information based on possibility distribution, *International Journal of General Systems*, vol. 9, pp. 43–58, 1983.

69. S. C. K. Shiu, X. Z. Wang, and D. S. Yeung, Neural-fuzzy approach for maintaining case-bases, in *Soft Computing in Case Based Reasoning* (S. K. Pal, T. S. Dillon, and D. S. Yeung, eds.), Springer-Verlag, London, pp. 259–273, 2001

70. Glass identification database, donated by Diagnostic Products Corporation, UCL Machine Learning Repository database, *http : //www1.ics.uci.edu/~mlearn/MLRepository.html*.

71. R. K. De and S. K. Pal, A connectionist model for selection of cases, *Information Sciences*, vol. 132, pp. 179–194, 2001.

72. S. K. Pal and P. K. Pramanik, Fuzzy measures in determining seed points in clustering, *Pattern Recognition Letters*, vol. 4, pp. 159–164, 1986.

73. Y. Hayashi, A neural expert system with automated extraction of fuzzy if–then rules and its application to medical diagnosis, in *Advances in Neural Information Processing Systems* (R. P. Lippmann, J. E. Moody, and D. S. Touretzky, eds.), Morgan Kaufmann, San Francisco, pp. 578–584, 1991.

74. Z. Pawlak, *Rough Sets: Theoretical Aspects of Reasoning about Data*, Kluwer Academic, Dordrecht, The Netherlands, 1991.

75. A. Skowron and C. Rauszer, The discernibility matrices and functions in information systems, in *Intelligent Decision Support: Handbook of Applications and Advances of the Rough Sets Theory* (R. Slowinski, ed.), Kluwer Academic, Dordrecht, The Netherlands, pp. 331–363, 1992.

76. S. K. Pal and A. Skowron (eds.), Special issue on rough sets, pattern recognition and data mining, *Pattern Recognition Letters*, vol. 24, no. 6, 2003.

77. T. Kohonen, *Self-Organizing Maps*, Springer Series in Information Sciences, Springer-Verlag, Heidelberg, 2001.

78. A. Roy and S. K. Pal, Fuzzy discretization of feature space for a rough set classifier, *Pattern Recognition Letters*, vol. 24, no. 6, pp. 895–902, 2003.

79. P. Mitra, S. K. Pal, and M. A. Siddiqi, Non-convex clustering using expectation maximization algorithm with rough set initialization, *Pattern Recognition Letters*, vol. 24, no. 6, pp. 863–873, 2003.

80. S. K. Pal, A. Ghosh, and B. Uma Shankar, Segmentation of remotely sensed images with fuzzy threshold, and quantitative evaluation, *International Journal of Remote Sensing*, vol. 21, no. 11, pp. 2269–2300, 2000.

CHAPTER 4

CASE ADAPTATION

4.1 INTRODUCTION

The first step in solving a new problem using a CBR system, as described in Chapter 3, is retrieving the most similar case(s) from the case base. Assuming that similar problems have similar solutions, the case(s) retrieved are used to derive the solution for the new problem. Usually, the past solution needs adjustment to fit the new situation. The process of fixing (i.e., adapting) the old solution is called *case adaptation*. The knowledge "repaired" to carry out the adaptation is referred to as *adaptation knowledge*.

There are two ways to acquire adaptation knowledge. The traditional approach is by interviewing domain experts and coding the task-specific adaptation knowledge manually into the CBR system. This knowledge may be represented as a decision table, semantic tree, or IF–THEN rules. Alternatively, the adaptation knowledge can be learned from the cases using machine-learning techniques. Through learning we generate specialized heuristics that relate the differences in the input specifications (i.e., problem attributes) to the differences in the output specifications (i.e., solution attributes). These heuristics can be used to determine the amount of adaptation that is suitable.

Acquiring adaptation knowledge through interviews with domain experts is both labor intensive and time consuming. It is also difficult to maintain the adaptation knowledge that has been acquired. Many application systems developed in the past, such as Chef, Judge, Clavier, and Juliana [1], employed this approach.

Foundations of Soft Case-Based Reasoning. By Sankar K. Pal and Simon C. K. Shiu
ISBN 0-471-08635-5 Copyright © 2004 John Wiley & Sons, Inc.

Recently, due to the availability of cases and the increase in computer processing power, many machine-learning approaches for deriving adaptation knowledge are developed. Among these, soft computing techniques play a unique role in capturing adaptation knowledge that is imprecise, uncertain, and approximate in nature.

In this chapter we describe a variety of ways of using soft computing techniques for case adaptation. We also provide a review of adaptation strategies and explain the traditional approaches of representing and using adaptation knowledge. Section 4.2 presents the traditional case adaptation strategies. In this section, three methods are described: reinstantiation, substitution, and transformation. These are followed in Section 4.3 by a brief description of a set of methodologies developed mostly by four leading groups using adaptation matrices, configuration techniques, the process of learning adaptation cases, and integrating rule- and case-based adaptation approaches. In Section 4.4 we explain adaptation methods based on various soft computing paradigms. Several techniques, including fuzzy decision trees, back-propagation neural networks, Bayesian models, support vector machines, and genetic algorithms, are explored for learning and optimizing the adaptation knowledge.

4.2 TRADITIONAL CASE ADAPTATION STRATEGIES

In this section we introduce three types of traditional case adaptation strategies:

- *Reinstantiation* is the simplest form of adaptation, in which the solution of the new problem is simply copied from the case retrieved and used directly, without modification.
- *Substitution* replaces parts of the old solution attributes that are invalid because they conflict with or contradict the new problem requirements. For example, in medical diagnosis, parts of a drug prescription in the past may need to be replaced and updated with more effective medicine for the treatment of new types of illnesses.
- *Transformation* is used when no appropriate substitute item is available. A tailored solution will be derived based on the constraints and the character-istics of the required solution. A constraint describes or defines the properties of a component(s) of the solution. It specifies what properties the solution should or should not have. The solution component must conform to the constraints, and no contradiction or conflict is allowed. To identify the constraints, some predefined expert knowledge or heuristic must be available.

In the following subsections we provide examples to illustrate three approaches. Table 4.1 compares the three methods. Both substitution and transformation are triggered or based on constraints and feedbacks from the problem and solution characteristics, while reinstantiation can be performed simply by applying the old solution directly.

TABLE 4.1   Characteristics of the Adaptation Strategy

Adaptation Strategy	Based on/Triggered by:
Reinstantiation	Null
Substitution	Constraints
	Feedback
Transformation	Constraints
	Feedback

4.2.1   Reinstantiation

Reinstantiation involves direct copying and use of an old solution from a retrieved case. If the similarity between the present and retrieved cases is high and no constraints or requirements are imposed on the solution required, a reinstantiation strategy can be used. The advantages of this method are its low cost of computation and quick response for users. Figure 4.1 provides an example of car fault diagnosis using a reinstantiation strategy. Case A is a new problem, and case B is the closest case found in the case library. Since these two cases are very similar and no constraint is being specified, the case B solution is copied and used to solve the case A problem.

In applying reinstantiation, the similarity measure is the most important factor. If the similarity is higher than a certain threshold, the previous successful solution will be copied and applied directly. Therefore, it is important to determine a suitable similarity measure and an appropriate threshold. In Chapter 3, several similarity measures, both with and without fuzziness, are described. They can be used here for this purpose. Following are the steps in reinstantiation.

Step 1. Determine a suitable similarity measure and an appropriate threshold.
Step 2. Retrieve the most similar case.

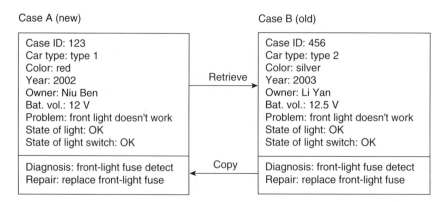

Figure 4.1   Reinstantiation.

Step 3. If the case retrieved has a similarity value above a specified threshold and no extra constraints are imposed on the solution required, the old solution is copied and applied to the new problem.

4.2.2 Substitution

Substitution replaces some parts of the old solution attributes that are invalid because they conflict with or contradict the new problem requirements. There are two types of substitution: constraint-based and feedback-based.

4.2.2.1 *Constraint-Based Substitution* A constraint, described or defined by the properties of a component(s) of the solution required, can generally be expressed by using such knowledge structures as a decision table, semantic tree, or IF–THEN rules. Therefore, in constraint-based substitution, the substituting components can be found by searching the corresponding knowledge structure. This knowledge structure is usually defined by experts through interviews. For example, in a semantic tree, the replacing items can be located by traversing the nodes. If these items are suitable for substitution, the contradicting part(s) of the old solution will be replaced (i.e., adapted) accordingly. Next, we provide an example to illustrate the idea.

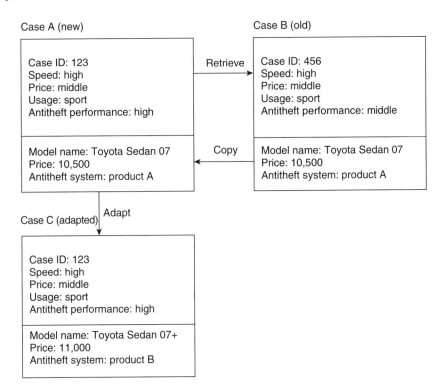

Figure 4.2 Constraint-based substitution.

Figure 4.2 illustrates a car-buying example. The objective of the system is to recommend a suitable car with the antitheft system required. Among all the product specifications, the customer specifies that the antitheft system satisfy the constraint "high performance." However, after comparing all the cases in the case library, the most similar case is case B, which has only a middle level of antitheft performance. In this situation, a constraint-based substitution will be carried out. The substituting value (i.e., an antitheft system that has a high level of performance and can be installed in a Toyota Sedan 07) can be found by searching a semantic tree. Figure 4.3 depicts the semantic tree of an auto system. The thick solid line, the thin solid line, and the dashed line represent the PART_OF, KIND_OF, and INSTANCE_OF semantic relation, respectively. For example, the security system is part of the auto system, antitheft system X is a kind of security system, and product A is an instance of antitheft system X.

First, the substitution procedure will check all other instances in the class antitheft system X in which product A belongs. Since no record is found, the search moves to the next class (i.e., antitheft system Y) to find a suitable substitution. In this class, products B and C are found. Both of them will be checked to ascertain whether they are of high performance and can be installed in Toyota Sedan 07. After checking, it is found that product B satisfies the constraint, and therefore it is used in the substitution. As a result, a new solution is derived based on a constraint-based substitution. Note that other components, such as the price of the new model, may need readjustment.

Following are the steps in constraint-based substitution.

Step 1. Retrieve the most similar case from the case base.

Step 2. Determine any constraint violation.

Step 3. Substitute the components violated using the predefined semantic knowledge.

Step 4. Perform the substitution and make other adjustments, if necessary.

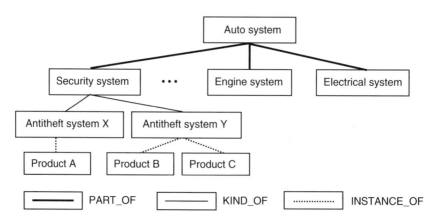

Figure 4.3   Semantic tree of an auto system.

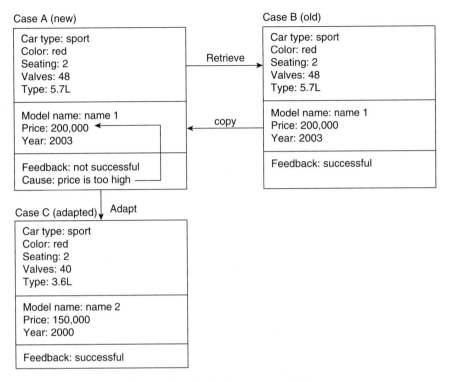

Figure 4.4 Feedback-based substitution.

4.2.2.2 Feedback-Based Substitution In feedback-based substitution, solution adaptation is an iterative process that is guided by evaluation of user feedbacks. Each time after a solution is carried out, feedback on its successfulness is returned. If the solution fails, the cause of the failure is analyzed and the result will be used to modify the adaptation process. Figure 4.4 gives an example of how to carry out feedback-based substitution. Case A is the query from the user and case B is the closest-matched case that is retrieved from the case library. The solution of case B is presented to the customer, but the feedback from the customer is: "Price is too high." In this situation, substitution will be carried out according to this information. The system will analyze the feedback and propose a solution that the customer will accept: that is, an alternative model of car and price will be proposed that satisfies the customer.

One of the advantages of feedback-based substitution is that the adaptation process is carried out interactively with user feedbacks. The solutions proposed are based on the needs and requirements of the users, and therefore they can be accepted more naturally. Following are the steps in feedback-based substitution.

Step 1. Retrieve the most similar case(s) from the case base.

Step 2. Collect the feedback from the user on the old solution.

Step 3. Analyze the outcome and determine the cause if the solution fails.

Step 4. Perform the substitution according to the feedback, and make other adjustments if necessary.

4.2.3 Transformation

Transformation is used when there is no appropriate substitute is available. A newly tailored solution will be derived based on the constraints and characteristics of the solution required. For example, sometimes it is difficult to find a substituting item that satisfies the problem requirements exactly. In this situation, the old solution will need to be adjusted, either partly or totally. Figure 4.5 gives an example of transformation adaptation. In the car-buying problem, one of the requirements of the user is the installation of a CD audio system in his new car. However, the closest-matched cases (i.e., cars) are all equipped with a tape system only, and after checking the semantic tree shown in Figure 4.6, no substitute is found to be available for these cars. To provide a workable solution, a tailored audio system with CD player is needed. The recorder component is removed from audio system A in the old solution, and a CD player is installed as a transformed solution.

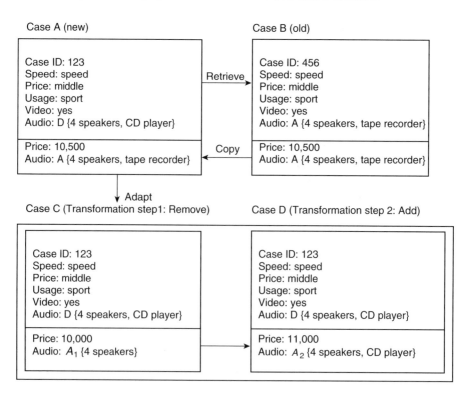

Figure 4.5   Transformation.

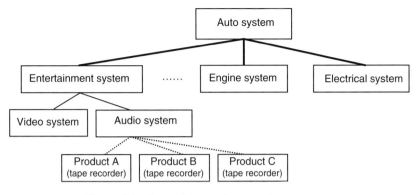

Figure 4.6   Semantic tree of a second auto system.

Following are the steps in transformation adaptation.

Step 1. Retrieve the most similar case from the case base.

Step 2. Repair the old solution by checking the semantic structure for available substitutions.

Step 3. If no substitution is available, transform the old solution by replacing some of its parts with suitable components.

Step 4. Add the new solution to the semantic structure for later use.

4.2.4   Example of Adaptation Knowledge in Pseudocode

In Sections 4.2.1 to 4.2.3, traditional adaptation strategies are presented. Here, a detailed example of a computer program that can be used to carry out the adaptation is given in pseudocode. For simplicity, a constraint-based substitution strategy (Section 4.2.2.1) is used. The adaptation knowledge is represented as IF–THEN rules, where the condition part is used for constraint checking and the decision part explains the actions to be performed. Figure 4.7 is an example of pseudocode that describes the adaptation process. It is written in C++ with comments separated by the "//" symbol.

4.3   SOME CASE ADAPTATION METHODS

Let us describe here some methodologies based on the aforesaid strategies, either individually or in combination. These methods are developed by four leading groups working on case adaptation. The first one [2] involves learning adaptation knowledge from the case data. This knowledge is stored as a set of "adaptation cases" and used to carry out the task of adaptation. The second method of Leake et al. [3] integrates the use of rule- and case-based reasoning for adaptation. Whereas the approach of Göker [4] uses an adaptation matrix to compute the

```
// Here the program begins.
```

...

```
// The classes CAntitheftsystemX and CAntitheftsystemY are part of the semantic tree. They are
   derived from the base class CSecuritysystem. The following declaration reveals their
   relationships.
class CAntitheftsystemX:CSecuritysystem  // In the semantic tree, CantitheftsystemX
is derived from CSecuritysystem, as shown in Figure 4.3.
{
...
CString Productname;
CString AntitheftPerformance;  // The value of this attribute may be "low", "middle,"
or "high."
float price;  // This is the price of an instance (e.g., product A) of class CAntitheftsystemX.
}productA;  // product A is an instance of class CAntitheftsystemX.

class CAntitheftsystemY:CSecuritysystem  // Declaration similar to the above.
{
...
CString Productname;
CString AntitheftPerformance;
float price;
}productB, productC;  // product B and product C are two instances of CantitheftsystemY.
They are brother nodes.


// Below is the declaration of the class Case.
class CCase
{
public:
int CaseID;  // This is the ID number of the Case.
CString Speed;  // The value may be "low," "middle," or "high."
CString Carprice;  // The value may be "low," "middle," or "high."
CString Usage;  // The value may be "sport," "commercial transport," or "family use."
CString AntitheftPerformance;  // The value may be "low," "middle," or "high."

float  Car_Price;  // This is the car price in numeric values.
CString Antitheftsystem;  // The value may be "productA," "productB," or "productC."

CCase* m_similar;  // The value is a pointer to the most similar case.
```

Figure 4.7 Pseudocode for constraint-based substitution.

```
CObList instancespointers;    // It contains the pointers to the instances of a class.
POSITION pos;  // It is used as an index parameter when traversing the instances pointers.
CArray<CString, CString> childrenclasses;  // It stores the names of the children
classes of a  base class.
public:
...
void getquery();  // It gets the query problem from the user interface.


CCase* retrievesimilar();  // It returns the pointer to the most similar case.
void copysolution(CCase* similar);  // It copies the old solution into the solution part of
 the query case, which will be adapted later.


void substitute();  // It carries out the substitution operation.


CString isinstancesof(CString instancename);  // Given the value of the
antitheftsystem, for example, "productA," it returns the name of the class AntitheftsystemX that it
 is instantiated  from.


void findinstancesofclass(CString classname);  // Given a class name, the pointers
to its instances are returned and stored into the variable CObList instancespointers. The list is then
traversed to locate an appropriate substituting item. For example, if classname is "CantitheftsystemY,"
then the CObList instancespointers will contain the pointers to products B and C.


CString findparentclassname(CString childclassname);  // Given the name of a
class, it returns the name of its parent class. It is applied when traversing a semantic tree.


void findchildrenclassname(CString parentclassname);  // Reversely, given the
name of a class, this function returns the names of its children classes and stores their names into the
Array<CString, CString> children classes.


void replace(CString classname);  // Given a class name, it searches its instances for a
suitable substituting item and then deploys this item to replace the invalid attribute in the old solution.
...
}


...
```

Figure 4.7   (*Continued*)

```
// Below is the process of substitution.
void CCase::substitute()
{
    ...
```

`CSecurity* p = locateobject(Antitheftsystem);` // Locate the object in the old solution. For example, if the value of Antitheftsystem is "product A," the pointer to the object product A is returned.

`if (p→Antitheftperformance.compare(Antitheftperformance)!=0)` // This is the condition part of the adaptation rule. It compares the value of the Antitheftperformance in the problem constraint with that of the antitheft product in the solution of the query case. If they are different, the constraint is identified as violated and the product must be substituted by carrying out the following decision part.

```
{
```

// First search the local branch where the object with the name Antitheftsystem inhabits.

`CString classname = isinstanceof(Antitheftsystem);` // Given the value of Antitheftsystem in the old solution, e.g., "productA," the name of the class that it is instantiated from, that is, "CantitheftsystemX," is returned.

`int flag=0;` // The flag is set to be 1 if substitution is performed successfully.

`replace(classname);` // Search the instances of the class with classname to find a substituting item and use it for substitution. Here the classname is CantitheftsystemX, corresponding to its instance, productA. Moreover, if the substitution is successful, the flag is set to 1.

`clearmemory();` // Clear the memory.

`if(flag==0)` // If the substitution item fails in the local search, go to other branches of the semantic tree.

```
{
```

`CString parentname = findparentclassname(classname);` // Get the name of the parent class of the class with classname. For example, if the classname is CantitheftsystemX, the parentname will be CSecuritysystem.

`findchildrenclassname(parentname);` // Find the names of other children classes of the class with parentname and store the names in the array children class. For example, if the parentname is "Csecuritysystem," then the name "CAntheftsystemY" is stored.

`for(count=0; count<childrenclasses.getsize();count++)` // For each children class, check its instances to find a suitable substituting item.

Figure 4.7 (*Continued*)

```
{
replace(childrenclasses[count]);
```
// If substitution is performed successfully, the flag is
set to 1.
```
if(flag==1) break;
}
clearmemory();
}}}
```
```
replace(CString classname)
```
// This is the procedure for searching a substituting item
among the instances of the class with classname and using the found item for substitution.
```
{
findinstancesofclass(classname)
```
// Find all the instances of a class with classname and
store the pointers into the CObList instance pointers.

```
CSecuritysystem* p = instancespointers.gethead();
```
// Get the pointer to the first
instance. For example, it is a pointer to product C.
```
for(pos = instancespointers.getheadposition(); pos!=NULL)
{
  if(p→Antitheftperformance.compare(Antitheftperformance)!=0)
  p = childrenpointers.getnext(pos);
```
// If the antitheft performance of the instance still
contradicts the query constraint, get the next instance.
```
   else
```
// If the instance satisfies the constraint, do the following.
```
   {
    CSecuritysystem* p1= locateobject(Antitheftsystem);
```
// For example, if the
antitheft system is "productA," the pointer to the object productA is returned. p1 will be used later
to adjust the car price.
```
    Antitheftsystem = p→Productname;
```
// Copy the name of the instance into the query
solution.
```
    Car_price = Car_price-p1→price+p→price;
```
// Adjust the car price because the
   Configuration of the antitheft system has been changed. p1→ price is the price of the antitheft
   system contradicting the constraint.
```
    instancespointers.removeall();
```
// Clear the memory.
```
    flag=1;
```
// Set the flag to 1 so that the substitution process is completed.
```
   }
}
else cout << "The old solution satisfies the constraint, no
adaptation is required.\n ";
```
// If the constraint is satisfied, show the above message.

Figure 4.7 (*Continued*)

```
if(flag==0)
cout<<"No suitable substituting item can be found in the semantic
tree.\n";
}
```

...

```
void main()   // This is the entry of the program.
{
CCase* query = new CCase;   // Create a new case pointer to hold the query.
query→getquery();   // Get the query case from the user interface.
query→m_similar = query->getsimilar();   // Get the case most similar to the query
case.
query→copysolution(m_similar);   // copy the old solution from the most similar case.
query→substitute();   // Perform substitution.
}
```

// Here the program ends.

Figure 4.7   (*Continued*)

necessary adaptation for attributes with numeric values, the one developed in [5] performs case adaptation based on configuration techniques. These are explained below.

4.3.1   Learning Adaptation Cases

An approach to acquiring adaptation knowledge automatically has been developed by Jarmulak et al. [2]. This method is suitable for configuration and formulation tasks, where the final solution consists of a constant number of components with relatively limited interactions between them. This type of problem can be regarded as a subclass of design problems. The adaptation process is as follows: After retrieving the most similar case(s) from the case base, the present case and the cases retrieved are analyzed and an "adaptation case" is constructed. For numerical attributes, the problem/solution differences between the current case and the cases retrieved are stored in the adaptation case. For nominal attributes, their "goodness" for predicting a proposed solution is stored. All the adaptation cases constitute what is called *adaptation knowledge*. Given a new current case, the most similar case(s) are retrieved from the case base. The adaptation cases are then used to compute the adjustment required for the solution proposed. This method has been tested successfully on the problem of designing tablet formulation/composition, which requires the selection of a set of excipients compatible with the drug, and at the same time satisfying a number of constraints.

4.3.2 Integrating Rule- and Case-Based Adaptation Approaches

A hybrid method described by Leake et al. [3] for case adaptation uses a combination of rule- and case-based reasoning. When a new problem case is encountered by the CBR system, the most similar case(s) are retrieved. The adaptation process begins with a small set of abstract transformation rules (see Section 4.2.3). The system then searches the case library and finds the information needed to operationalize (i.e., firing the rules) the transformation rule and apply it to the problems at hand. The system improves its adaptation capability by case-based reasoning applied to the case adaptation process itself: A trace of the problem-solving steps of a proposed solution is stored and reused when similar problems arise in the future. If the new solution is still not acceptable after using both rules and cases, users can perform a specific/tailored adaptation through use of the system's user interface. The system consists of the following components for adaptation:

- *Rule-based (transformation rules) adaptation component.* The system performs an adaptation by using adaptation rules, then applies it to the problems at hand. These rules are obtained by interviewing domain experts.
- *Case-based adaptation component.* If no rules are suitable for use, the system tries to retrieve an adaptation case describing successful earlier adaptation of a similar problem. If the adaptation case is found, it is used for the adaptation; otherwise, the manual adaptation process is triggered.
- *Manual adaptation component.* If rule- and case-based adaptations fail to generate an acceptable solution, the user can perform the adaptation manually through the system's user interface.

After the adaptation is successful, the adapted solution and the problem-solving steps (i.e., adaptation case) are stored for future use.

4.3.3 Using an Adaptation Matrix

In problems where the features are having interactions (e.g., if there is a dependency relationship between two features), the adaptation process needs to consider these interactions carefully. In Göker [4] an adaptation matrix is developed to describe this characteristic. The idea is summarized as follows: The attributes of an object (i.e., a case) are classified into two groups: the independent (base) attributes and the dependent (derived) attributes. For example, a dependent attribute may be the price of a product, while the independent attributes may be the components needed in the manufacturing process. An adaptation matrix can be developed that describes the relationship (or influence levels) of the independent attributes with (on) the dependent attributes. Using this matrix, the effects on dependent attributes of changes in the independent attributes can be computed. Given a new case, the system retrieves the most similar case(s) from the case base. The differences between the present case and the case(s) retrieved are checked against the adaptation matrix. Adaptation of the retrieved solution can thus be performed. For example, only limited adaptation is required if changes are made to independent attributes that have minimal effects on the dependent attribute (e.g., the solution). On the other hand, if changes are made to the independent attributes that are having substantial effects on the dependent attribute, the solution may need a large adjustment.

4.3.4   Using Configuration Techniques

An adaptation method, based on configuration techniques is described by Wilke et al. [5]. It uses an object-oriented model to facilitate the adaptation process. A case is an instance of a class (or a concept). It has a set of attributes that characterize its properties. The relationships among these classes (i.e., concepts), such as KIND_OF, PART_OF and HAS_A, are used to build a hierarchical structure that can provide information for the adaptation. The adaptation tries to build a complete solution by configurating the components contained in the cases retrieved. The configuration process starts by changing the major components, then subcomponents, and so on. In each process, a set of constraints that is imposed on the solution is used to evaluate the result of each configuration. The object hierarchy serves as semantic knowledge of the relationships among the components that are being changed or substituted for in the configuration process. This technique has been used in assisting the configuration of computer systems that are sold on the Web.

Other alternative ways of case adaptation include the divide-and-conquer approach developed by Fuchs and Mille [6], in which a problem is decomposed into subproblems. For each subproblem a corresponding adaptation task is performed. By combining the individual adapted components, the overall solution is obtained. A structural adaptation method that is based on evaluation of the structural similarity (e.g., class–subclass relationship) among cases is reported [7]. When performing the adaptation, both the values of the attributes and the structural similarity are taken into consideration. The approach based on introspective reasoning [8], involves a feedback mechanism to iteratively fine-tune the adaptation process until the user is satisfied with the solution. The use of the idea of replaying the problem-solving steps for adaptation is available [9].

4.4   CASE ADAPTATION THROUGH MACHINE LEARNING

In previous sections we have described three traditional case adaptation strategies and some methods that are based on them, either individually or in combination. The adaptation knowledge is derived primarily by interviewing the domain experts. This process of acquiring adaptation knowledge is both labor intensive and time consuming. The knowledge is also difficult to maintain. In this section, several machine learning techniques, including those based on fuzzy decision trees, artificial neural networks, Bayesian models, support vector machines, and genetic algorithms are described. The task of learning is to generate specialized heuristics that relate the differences in the input specifications (i.e., problem attributes) to the differences in the output specifications (i.e., solution attributes). These heuristics can be used to determine the suitable amount of adaptation that is required. One of the advantages of these techniques is that the adaptation knowledge can be learned automatically from the cases, and therefore this knowledge is more robust. Furthermore, maintenance of this knowledge can be better controlled and managed by relearning and retraining with the new cases. The idea is expressed below.

First, let there be M clusters in the case base obtained through the use of clustering techniques described in Chapter 3. For a cluster LE $= \{e_1, e_2, \ldots, e_m\}$, let each case be denoted as $e_i = (x_{i1}, x_{i2}, \ldots, x_{in}, \theta_i)$ for $i = 1, 2, \ldots, m$, where the jth attribute x_{ij} denotes the value of feature $F_j (1 \leq j \leq n)$ of the ith case, and θ_i corresponds to the action or value of the solution of the ith case. Compute, for every case e_k in the cluster LE,

$$\phi_{ik} = e_i - e_k = (x_{i1} - x_{k1}, x_{i2} - x_{k2}, \ldots, x_{in} - x_{kn}, \theta_i - \theta_k),$$

where ϕ_{ik} is called the ith discrepancy vector of the kth case, and $i, k = 1, 2, \ldots, m$. For simplicity, we denote

$$X_i = (x_{i1}, x_{i2}, \ldots, x_{in}),$$
$$\Delta X_{ik} = (x_{i1} - x_{k1}, x_{i2} - x_{k2}, \ldots, x_{in} - x_{kn}),$$
$$\Delta \theta_{ik} = (\theta_i - \theta_k)$$

Given a query problem X (i.e., the problem part of a case), the system retrieves the most similar case $e_0 = (X_0, \theta_0)$. To determine the amount of adjustment (i.e., adaptation) needed to modify the solution θ_0, the discrepancy vectors are used as training examples to generate some specialized heuristics (knowledge in the form of rule, say) that relate differences in the input specifications (i.e., problem attributes) to differences in the output specifications (i.e., solution attributes).

The discrepancy vectors, which are used for the learning of adaptation knowledge, are defined as

$$\phi_{i0} = e_i - e_0 = (\Delta X_{i0}, \Delta \theta_{i0}) \tag{4.1}$$

where $e_i = (X_i, \theta_i)$ denotes the neighboring cases of e_0. After training, the corresponding adjustment amount $\Delta \theta_0$ can be determined. Thus, the new adapted solution for case X_0 is obtained (i.e., $\theta = \theta_0 + \Delta \theta_0$). In the following sections we describe several soft computing techniques for learning this adaptation knowledge.

4.4.1 Fuzzy Decision Tree

A fuzzy decision tree method such as fuzzy ID3 has been described in Chapter 3 for classifying cases. Fuzzy decision trees can also be used to represent adaptation knowledge. Next, we provide the steps for acquiring such knowledge.

Step 1. Learn the feature weights of the problem attributes.
Step 2. Cluster the cases using the weighted distances.

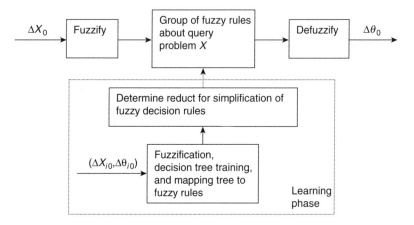

Figure 4.8   Adaptation using fuzzy decision tree.

Step 3. For every case in each cluster, compare the differences between this case and other cases in both problem attributes and solution attributes. Obtain a set of discrepancy vectors.

Step 4. Fuzzify the discrepancy vectors.

Step 5. For every case, a fuzzy decision tree is obtained through training using the fuzzy discrepancy vectors. Generate the fuzzy decision rules from the tree, and simplify the rules using the rough set technique (i.e., reducts).

Step 6. Given a new problem, retrieve the most similar case from the case base. Fuzzify the difference between these two cases in the problem parts. Use the fuzzified difference as input to the fuzzy decision tree. Obtain the adjustment in the form of a linguistic variable with certainty factors. Defuzzify them to get the adjustment amount.

Step 7. Use the adjustment values to adapt the old solution.

The steps above are summarized in Figure 4.8. Note that in step 1, the weights of the attribute features can be learned using the algorithms described in Section 3.5. For step 2, the case clustering, the k-NN approach in Section 3.2 or the fuzzy-c means approach in Section 3.4 can be used. For step 4, equation (2.10) can be applied to generate the fuzzy membership functions. The defuzzification operation can be carried out using equation (5.13).

4.4.2   Back-Propagation Neural Network

Many neural network models, such as the radial basis function (RBF) neural network [10] and the back-propagation (BP) neural network, can be used to acquire the adaptation knowledge using discrepancy vectors. Here we introduce use of a BP network for such a task. (For details on the BP and RBF neural networks and their training algorithms, readers can refer to Appendix B.) Following are the steps in using a back-propagation (BP) neural network for case adaptation.

Step 1. Learn the feature weights of the problem attributes.

Step 2. Cluster the cases using the weighted distances.

Step 3. For every case in each cluster, compare the differences between this case and other cases in both problem attributes and solution attributes. Obtain a set of discrepancy vectors.

Step 4. Preprocess (e.g., normalize) the discrepancy vectors for training the neural network.

Step 5. For each case, use its discrepancy vectors to train the neural network.

Step 6. Given a new problem, retrieve the most similar case from the case base. Compute the difference between their problem attributes. Preprocess this difference and use it as the input to the neural network.

Step 7. Obtain the adjustment amount from the neural network by inputting the preprocessed differences of the attribute values.

Step 8. Postprocess (e.g., denormalize) the adjustment amount and use it to adapt the old solution for the query problem.

The steps above are summarized in Figure 4.9.

4.4.3 Bayesian Model

We discussed the Bayesian model in Section 2.4.2 for case indexing. This model can also be used in case adaptation if the sample space contains fuzzified discrepancy vectors. Given a new query, the most similar case is retrieved and the difference between their problem parts is input to the Bayesian model. The output is the class label that identifies the adjustment strategy. The following steps describe how to adapt the old solution using the Bayesian model.

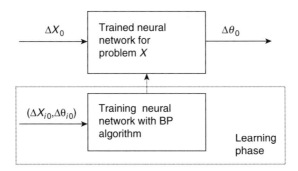

Figure 4.9 Adaptation using BP neural network.

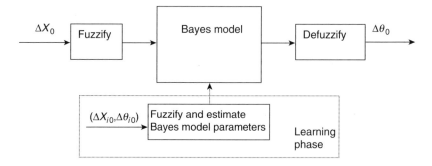

Figure 4.10   Adaptation using the Bayesian model.

Step 1. Learn the feature weights of the problem attributes.

Step 2. Cluster the cases using the weighted distances.

Step 3. For every case in each cluster, compare the differences between this case and other cases in both problem attributes and solution attributes. Obtain a set of discrepancy vectors.

Step 4. Fuzzify the discrepancy vectors. Estimate the parameters of the Bayesian model.

Step 5. Given a new problem, retrieve the most similar case from the case base. Fuzzify the difference between these two cases in the problem parts.

Step 6. Input the problem difference into the Bayesian classifier. Identify the class label for adjustment using the Bayes formula.

Step 7. Obtain the adjustment in the form of a linguistic variable with certainty factors. Defuzzify them to get the adjustment amount.

Step 8. Use the adjustment amounts to adapt the old solution.

The steps above are summarized in Figure 4.10.

4.4.4   Support Vector Machine

Support vector machines (SVMs) [11,12] are a class of classifiers that provide good generalization (i.e., can achieve high accuracy of classification even with a relatively small number of training samples). Therefore, if the available cases are limited, this method can be used as an alternative to the more traditional approaches, such as the Bayesian model and the BP neural networks. The process of using the support vector machine in the case adaptation is as follows. Given a query problem X (i.e., the problem part of a case), the system retrieves the most similar case $e_0 = (X_0, \theta_0)$. The discrepancy vectors $(\Delta X_{i0}, \Delta \theta_{i0})$ with respect to X are generated. For simplicity, we assume that the values of $\Delta \theta_{i0}$ fall into only two categories, labeled $+1$ or -1, representing two different types of adaptations (e.g., with adaptation or without adaptation). To predict the value of $\Delta\theta_0$ given ΔX_0, we just need to identify its corresponding label. To achieve this goal, an optimal separating hyperplane is constructed using the discrepancy vectors in the form of $(\Delta X_{i0}, \text{label}_{i0})$, where label_{i0} is $+1$ or -1. After the hyperplane is constructed, ΔX_0 is input to the classifier and the corresponding $\Delta\theta_0$ is obtained by checking the output of the SVM.

In situations where the discrepancy vectors are not linearly separable, the sample space is transformed to a new space so that the data can be separated linearly. This is performed using a kernel mapping f such that

$$\Delta X_{i0} = (x_{i1} - x_{01}, \ldots, x_{in} - x_{0n}) \mapsto f(\Delta X_{i0}) = (f_1(\Delta X_{i0}), \ldots, f_N(\Delta X_{i0})) \quad (4.2)$$

Usually, more than one hyperplane can be constructed for correct classification in the sample space. For example, two hyperplanes L_2 (the solid lines in Figure 4.11) can be used to separate the samples. The optimal one should be chosen so as to have a lower possibility of misclassification on new (unknown) samples. Note that in Figure 4.11,

$$\begin{aligned} L_1 : \quad & Kf(X) + b = 1 \\ L_2 : \quad & Kf(X) + b = 0 \\ L_3 : \quad & Kf(X) + b = -1 \end{aligned} \quad (4.3)$$

where K and b are parameters determining the position of the hyperplanes.

We find that the hyperplane L_2 in Figure 4.11b has a better separation margin (ability) than the one in Figure 4.11a. This is because the distance between the two parallel lines L_1 and L_3 in Figure 4.11b is larger than that of Figure 4.11a, and this can reduce the possibility of misclassification on the test samples. The distance between the parallel lines, such as those in Figure 4.11, is $1/||K||$, where $||K|| = \left(\sum_{i=1}^{N} k_i^2\right)^{1/2}$ for $K = (k_1, k_2, \ldots, k_N)$. The best hyperplane can be found

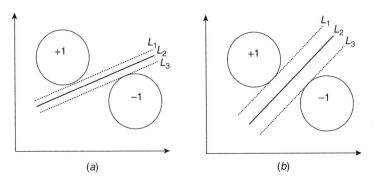

(a)   (b)

Figure 4.11   Samples separated with the hyperplanes.

by selecting an appropriate K that has the minimal $\|K\|$ value and satisfies the constraints in the following quadratic programming problem:

$$\text{Min}\langle K, K \rangle \qquad y_i(\langle Kf(X_i)\rangle + b) \geq 1, \qquad i = 1, 2, \ldots, l \qquad (4.4)$$

where $\langle \cdot, \cdot \rangle$ denotes the inner product, l the number of training samples, and y_i corresponds to label$_{i0}$. To find K and b, a Lagrangian function is constructed as

$$L(K, b, \alpha) = \tfrac{1}{2}\langle K, K \rangle - \sum_{i=1}^{l} \alpha_i[y_i(\langle K, f(X_i)\rangle + b) - 1], \qquad \alpha_i \geq 0 \qquad (4.5)$$

According to the Kuhn–Tucker saddle point condition [11,12], if there exists a point (K^*, b^*, α^*) satisfying

$$L(K^*, b^*, \alpha) \leq L(K^*, b^*, \alpha^*) \leq L(K, b, \alpha^*) \qquad (4.6)$$

the optimal solution $h^* = (K^*, b^*)$. To obtain (K^*, b^*, α^*), we need to get the minimum value of L with respect to K and b and the maximum value with respect to α. Let $\partial L/\partial K = \partial L/\partial b = 0$, we have $K^* = \sum_{i=1}^{l} \alpha_i^* y_i f(X_i)$ and $\sum_{i=1}^{l} \alpha_i^* y_i = 0$. By substituting them into L, K and b are eliminated from equation (4.5). The value of α^* can thus be achieved by solving a quadratic programming problem with respect to α^*. To obtain the value of b^*, a group of support vectors, which are on the margin hyperplanes, are identified using the Karush–Kuhn–Tucker condition [11,12]. These support vectors are then used to get the value of b^* from the equation of the hyperplanes (4.3).

Rewriting the optimal hyperplane in the form of inner product, we have

$$\sum_{i=1}^{l} \alpha_i^* y_i \langle f(X_i), f(X)\rangle + b^* = 0$$

It can also be rewritten in the form of kernel function $\sum_{i=1}^{l} \alpha_i^* y_i \, \text{Ker}(X_i, X_j) + b^* = 0$, where

$$\text{Ker}(X_i, X_j) = \langle f(X_i), f(X)\rangle \qquad (4.7)$$

Thus, a general form of the classifier can be represented by

$$\text{Class}(X) = \text{sgn}\left(\sum_{i=1}^{l} \alpha_i^* y_i \text{Ker}(X_i, X) + b^* \right) \qquad (4.8)$$

where

$$\text{sgn}(x) = \begin{cases} 1 & x > 0 \\ 0 & x = 0 \\ -1 & x < 0 \end{cases} \qquad (4.9)$$

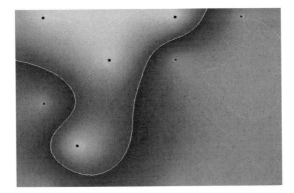

Figure 4.12   Binary classification of nonlinearly separable data using SVM.

Now, for case adaptation, given a new query problem X, we can input it into equation (4.8) and output the class label (i.e., either -1 or $+1$) to identify the corresponding adaptation strategy. The following kernel functions can be used.

$$\text{Ker}(X_i, X) = X_i^{\text{T}} X; \; \text{Ker}(X_i, X) = (X_i^{\text{T}} X + 1)^d; \; \text{Ker}(X_i, X) = \exp\left(\frac{-||X - X_k||^2}{\sigma^2}\right)$$

One of the advantages of SVM is that it requires relatively fewer samples to achieve a satisfactory result, as can be seen in Figure 4.12. The SVM classifier has a better performance than the BP neural networks, which require many more samples and computing efforts to reach the same level of accuracy.

For multiclass problems (i.e., the sample data are classified into n classes), Inoue and Abe [13] have proposed a fuzzy support vector machine to generate the optimal boundaries among multiclasses. The extension of SVM with fuzziness makes it more useful for case adaptation. The steps in using SVM for learning adaptation knowledge are as follows:

Step 1. Learn the feature weights of the problem attributes.

Step 2. Cluster the cases using weighted distances.

Step 3. For every case in each cluster, compare the differences between this case and other cases in both problem and solution attributes. Obtain a set of discrepancy vectors.

Step 4. For each case, construct the support vector machine using the discrepancy vectors.

Step 5. Given a new query, find its most similar case and compare the discrepancy in the problem part.

Step 6. Input ΔX_0 and output the label corresponding to $\Delta \theta_0$ to determine the adaptation strategy.

Step 7. Use the adaptation strategy to derive a new solution.

The steps above are summarized in Figure 4.13.

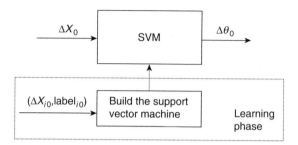

Figure 4.13 Adaptation using support vector machine.

4.4.5 Genetic Algorithms

In previous algorithms, adaptation knowledge has been acquired directly from cases. However, in situations when there are not enough cases, we can use genetic algorithms. The idea is that when an old solution is retrieved, a modification process is initialized randomly. The solution adapted is tested and the feedback is collected to determine its fitness. This process is repeated for many cycles until a satisfactory solution is obtained. The modification can be expressed as a vector and encoded as a chromosome. Genetic operators such as selection, crossover, and mutation can be used to work on these vectors and produce offspring. The evaluation of the fitness value of the adapted solution can be determined either by a domain-specific model or by tests carried out in the real world. For example, in planning and synthesizing tasks where a CBR system is used to assist a planner or architect to construct an artifact based on the matching against a set of prototypical artifacts in the case base, the adaptation involves the determination of the correct features in the correct places as well as in the correct order, the values corresponding to these parameters can be determined using GA.

To illustrate the principle above, the following example is used. In Table 4.2, for problem 1, three modification vectors are initialized randomly and used to adapt the

TABLE 4.2   GA for Case Adaptation

Problem	Modification Vectors	Fitness of the New Solution
1	$\Delta v_{11}^1, \Delta v_{12}^1, \ldots, \Delta v_{1n}^1$	fit_{11}
	$\Delta v_{21}^1, \Delta v_{22}^1, \ldots, \Delta v_{2n}^1$	fit_{12}
	$\Delta v_{31}^1, \Delta v_{32}^1, \ldots, \Delta v_{3n}^1$	fit_{13}
2	$\Delta v_{11}^2, \Delta v_{12}^2, \ldots, \Delta v_{1n}^2$	fit_{21}
	$\Delta v_{21}^2, \Delta v_{22}^2, \ldots, \Delta v_{2n}^2$	fit_{22}
	$\Delta v_{31}^2, \Delta v_{32}^2, \ldots, \Delta v_{3n}^2$	fit_{23}

old solution. Δv_{ij}^{k} is the jth adjustment value in the ith modification vector for the kth problem. fit_{ki} is the fitness value of the solution adapted with the ith adaptation vector for the kth problem. Each vector is coded and used as the chromosome in genetic operations. After carrying out the genetic operations successively, the modification vectors are optimized gradually. The optimized chromosome can then be used for the adaptation.

4.5 SUMMARY

In this chapter we have presented some of the traditional and the soft computing–based approaches for case adaptation. Traditionally, the adaptation knowledge is acquired through interviews with human experts and coded manually into a CBR system. Three types of traditional adaptation strategies are described: reinstantiation, substitution, and transformation. These methods are both labor intensive and time consuming. Maintenance of the knowledge base is also difficult. For machine learning–based methods, soft computing techniques, including those based on fuzzy decision trees, back-propagation neural networks, Bayesian models, support vector machines, and genetic algorithms, are presented. One of the advantages of these techniques is that adaptation knowledge can be determined automatically from the cases, so that this knowledge is more robust. Furthermore, maintenance of this knowledge base becomes easier by relearning or retraining with the new cases available. Although we have presented only the broad principles, the particular algorithms can be formulated depending on the problems at hand.

REFERENCES

1. J. L. Kolodner, *Case-Based Reasoning*, Morgan Kaufmann, San Francisco, 1993.

2. J. Jarmulak, S. Craw, and R. Rowe, Using case-base data to learn adaptation knowledge for design, in *Proceedings of the Seventeenth International Joint Conference on Artificial Intelligence (IJCAI-01)*, Seattle, WA, Morgan Kaufmann, San Francisco, pp. 1011–1016, 2001.

3. D. B. Leake, A. Kinley, and D. Wilson, Acquiring case-adaptation knowledge: a hybrid approach, in *Proceedings of the Thirteenth National Conference on Artificial Intelligence*, Menlo Park, CA, AAAI Press, Menlo Park, CA, pp. 684–689, 1996.

4. M. H. Göker, Attribute types and adaptation of technical objects, in *Proceedings of the Sixth German Workshop on Case-Based Reasoning (GWCBR-98)*, Universität Rostok, Berlin, vol. 7, pp. 169–178, 1998.

5. W. Wilke, B. Smith, and P. Cunningham, Using configuration techniques for adaptation, in *Case-Based Reasoning Technology from Foundations to Applications* (M. Lenz, B. Bartsch-Spörl, H. D. Burkhard, and S. Wess, eds.), Springer-Verlag, Berlin, pp. 139–168, 1998.

6. B. Fuchs and A. Mille, A knowledge-level task model of adaptation in CBR, in *Proceedings of the Third International Conference on Case-Based Reasoning (ICCBR-99)*, Seeon Monastery, Munich, Germany, Springer-Verlag, Berlin, pp. 25–27, 1999.

7. K. Börner, C. H. Coulon, E. Pippig, and E. C. Tammer, Structural similarity and adaptation, in *Proceedings of the Third European Workshop on Case-Based Reasoning (EWCBR-96)*, Lausanne, Switzerland, Springer-Verlag, Berlin, pp. 58–75, 1996.

8. D. B. Leake, Learning adaptation strategies by introspective reasoning, in *Proceedings of the AAAI-93 Workshop on Case-Based Reasoning*, Menlo Park, CA, AAAI Press, Menlo Park, CA, pp. 57–63, 1993.

9. W. Wilke and R. Bergmann, Techniques and knowledge used for adaptation during case-based problem solving, in *Proceedings of the Eleventh International Conference on Industrial and Engineering Applications of Artificial Intelligence and Expert System (IEA/AIE-98)*, Castellón, Spain, Springer-Verlag, Berlin, pp. 497–506, 1998.

10. T. Koiranen, T. Virkki-Hatakka, A. Kraslawski, and L. Nyström, Hybrid, fuzzy and neural adaptation in case-based reasoning system for process equipment selection, *Computer and Chemical Engineering*, vol. 22, pp. 997–1000, 1998.

11. B. Schölkopf and A. J. Smola, *Learning with Kernels-Support Vector Machines: Regularization, Optimization, and Beyond*, MIT Press, Cambridge, MA, 2002.

12. N. Cristianini and J. Shawe-Taylor, *An Introduction to Support Vector Machines and Other Kernel-Based Learning Methods*, Cambridge University Press, Cambridge, 2000.

13. T. Inoue and S. Abe, Fuzzy support vector machines for pattern classification, in *Proceedings of the International Joint Conference on Neural Networks (IJCNN-01)*, Washington, DC, IEEE Press, Piscataway, NJ, vol. 2, pp. 1449–1454, 2001.

CHAPTER 5

CASE-BASE MAINTENANCE

5.1 INTRODUCTION

In previous chapters we have described different tasks of CBR systems, such as selection and extraction of prototypical cases, retrieval of cases, establishing an appropriate similarity function, and defining various adaptation mechanisms. Since the CBR systems tend to grow in an evolutionary manner, constant maintenance of the knowledge is necessary to ensure correct system performance in response to changes in task or environment. The importance of maintaining traditional expert systems, in particular rule- and frame-based expert systems, has been investigated and well documented [1–7]. However, the issue of maintenance of CBR systems has largely been neglected in the past, and only recently, due to the increasing number of systems being used in the industry, has the problem of case-base maintenance (CBM) become more prevalent [8–20]. In essence, case-base maintenance can be viewed as a process of refining a CBR system to facilitate future reasoning for a particular set of performance objectives [21]. These objectives are usually defined by users depending on the task domain and external environment. There are two types of maintenance tasks: qualitative and quantitative. *Qualitative maintenance* deals with assurance of the correctness, consistency, and completeness of the CBR system; *quantitative maintenance* is concerned with assurance of the problem-solving efficiency (e.g., the average problem-solving time), the practical limit of the size of the case base (e.g., storage limits), the reorganization of the case indexes, the

Foundations of Soft Case-Based Reasoning. By Sankar K. Pal and Simon C. K. Shiu
ISBN 0-471-08635-5 Copyright © 2004 John Wiley & Sons, Inc.

changes of case representation structures, and other house keeping tasks. So far, many policies and techniques have been applied to maintaining case bases, some of them based on the soft computing paradigm [22,23]. In this chapter we provide a brief review of different methods of case-base maintenance, explain the significance of soft computing, and demonstrate some maintenance algorithms based on soft computing. In Section 5.2 the background of knowledge-based system maintenance is provided for the convenience of readers. The meaning of case-base maintenance, from both the qualitative and quantitative points of view, is explained in Section 5.3. In Sections 5.4 and 5.5, respectively, we describe in detail a rough-fuzzy hybridization method and a fuzzy integral–based competence model for case-base maintenance.

5.2   BACKGROUND

In the early 1990s there was an explosion of activities in the area of expert system maintenance [then known in the literature as validation and verification (V&V)]. For example, one of the longest-sequenced workshops at the AAAI (American Association for Artificial Intelligence) meeting has been the Workshop on Verification, Validation and Testing of Intelligent Systems. The first five workshops were held during 1988–1992, and the latest one was organized by O'Leary and Preece [24] in 1998. The International Joint Conferences on Artificial Intelligence (IJCAI) had workshops on V&V since 1989. The European Conference on AI (ECAI) also has a number of workshops on V&V. Special issues on verification and validation of expert systems are available in a number of journals: *International Journal of Human-Computer Studies*, vol. 44, no. 2, 1996; *International Journal of Intelligent Systems*, vol. 9, no. 8, 1994; *International Journal of Expert Systems*, vol. 6, no. 3, 1993; and *Expert Systems with Applications*, vol. 1, no. 3, 1990 and vol. 8, no. 3, 1995. Recently, research on maintenance and management of knowledge-based systems has become a part of the more general issue of knowledge management (KM). Various KM conferences and workshops have addressed the issue of maintenance of intelligent systems: the ECAI-02 workshop on knowledge transformation; the ECAI-02 workshop on knowledge management and organizational memories; the Second International Conference on Knowledge Management, I-KNOW-02; and the Fourth International Conference on Practical Aspects of Knowledge Management. Parallel to the research activities in the maintenance of rule- and frame-based knowledge-based systems, the maintenance of CBR systems was recognized in the late 1980s, and since then research has been conducted by a number of researchers, such as O'Leary [25], Racine et al. [26–29], Leake et al. [21,30–32], Smyth and McKenna [33–37], and Shiu et al. [38–40]. Recently, a special issue on maintaining CBR systems is published in *Computational Intelligence Journal*, vol. 17, no. 2, May 2001. However, the work done so far is quite diverse and without having a unified framework on case-base maintenance. In the next section a systematic method of defining case-base maintenance tasks is given based on its qualitative and quantitative aspects.

5.3   TYPES OF CASE-BASE MAINTENANCE

Case-base maintenance involves the policies and techniques of adding, deleting, and updating cases, indexes, and other knowledge structures (e.g., similarity and adaptation) in a CBR system in order to guarantee its ongoing effectiveness and performance. Its activities can be divided broadly into two categories: (1) qualitative maintenance and (2) quantitative maintenance. Qualitative maintenance refers to the assurance of the effectiveness of the CBR system, which includes the correctness, consistency, and completeness of the system. Quantitative maintenance aims at assuring the problem-solving efficiency of a CBR system, such as improving the average problem-solving time, controlling the size of the case base, and reorganizing the case index structure.

5.3.1   Qualitative Maintenance

The purpose of qualitative maintenance is to assure, mainly, the three characteristics of CBR systems: correctness, consistency, and completeness.

5.3.1.1   Correctness   The correctness of a CBR system is its ability to "solve" input query problems (i.e., whether a correct CBR system has been built). For example, in classification problems such as medical diagnosis and equipment failure analysis, a new case is matched against those in the case base to determine the correct class label. The correctness of the CBR system is therefore defined as the degree of successful classification of cases. On the other hand, if the solution component is a real value, such as the housing price or rental cost of apartments, the correctness of the CBR system could be defined as the percentage difference between the actual and predicted housing prices. However, in synthesis tasks where a CBR system is used to assist a planner or architect to construct an artifact based on matching against a set of prototypical artifacts in the case base, the correctness would be more difficult to define. This is because synthesizing tasks involve placing the correct features in the correct places as well as in the correct order. These tasks are more difficult to evaluate objectively. Determination of the correctness of a CBR system is application and domain dependent.

5.3.1.2   Consistency   The consistency of a CBR system involves its quality of providing solutions that are not in conflict with or contradict the same problem. In general, a CBR system is considered to be inconsistent if it contains one or more of the following types of cases in its case base: redundant, conflicting, alternative, erroneous, dead cases. Inconsistencies are anomalies and are considered as potential errors. Detection and analysis of these potential errors are crucial in building and maintaining CBR systems. The aforesaid types of problematic cases are described as follows:

1. *Redundant cases.* If two cases are the same (i.e., case duplication) or if one case subsumes another case, one of the cases duplicated or the case subsumed can be removed from the case base without affecting the overall problem- solving ability of the CBR system. The meaning of *subsumption* is as follows: Given two cases e_p and e_q, when case e_p subsumes case e_q, case e_p can be used to solve more problems than e_q. In this case, e_q is said to be *redundant*. For example, let e_p and e_q be two successful cases of bank credit card application. Let the "salary" of e_p be above 12K and the "salary" of e_q be above 20K, and other attributes be equal. If e_p gets credit card approval with a salary above 12K, e_q will get approval as well. This means that case e_p subsumes case e_q and e_p is a general case of e_q. Therefore e_q can be removed.

2. *Conflicting cases.* Conflicting cases are two or more cases that are very similar on every problem attribute, yet propose conflicting solutions. This conflict may be due to the erroneous input of case attributes or may be due to a change in the "correct" solution over time.

3. *Alternative cases.* In many applications, offering alternative solutions to the same problem is very useful (e.g., recommending various holiday packages to the tourists). Therefore, if a system is designed to offer alternative solutions rather than pure similarity-based suggestions, the detection of these alternative cases will be very useful.

4. *Erroneous cases.* A case that is inconsistent with the background knowledge of the system is an erroneous case. For example, if an attribute value or suggested solution to a case has something like "age of a customer is 2003," it is called an erroneous case because it does not match with reality (i.e., background knowledge). Erroneous cases result from errors introduced during the case capturing or entering process. These errors can be detected using standard data validation methods such as the check digit.

5. *Dead cases.* A case is a dead case if it cannot be retrieved from the case base due to missing important attributes or erroneous feature values. Given a new case, the similarity score with this dead case is always zero (i.e., totally dissimilar).

5.3.1.3 *Completeness*   Completeness of a CBR system involves its coverage of the problem space in the target domain (i.e., the CBR system contains all the essential cases that could be used to generate solutions to all possible current cases). In the process of building CBR systems, cases are collected incrementally, and the completeness of a case base is also evolving over time. However, the addition of cases may not necessarily improve the completeness of a case base; therefore, identification of the essential cases as well as "missing knowledge" becomes

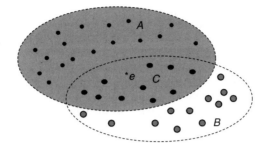

Figure 5.1   Case coverage and reachability.

the primary task in building a comprehensive CBR system with satisfactory completeness. There are two characteristics that could be used to describe the completeness of a CBR system: (1) coverage of cases and (2) reachability of cases [33,34]. *Coverage* of a case is the set of target problems (i.e., cases) that this case in question can be used to solve. The *reachability* of a target problem (i.e., a case), on the other hand, is the set of all cases that can be used to solve it. These concepts are illustrated in Figure 5.1. Here sets A and B represent the coverage and reachability of case e, respectively. Set C, which is the intersection of A and B, represents those cases that can be solved by case e as well as can be used to solve case e. Note that the cases with larger coverage have higher contributions to the competence of (i.e., the range of problems that can be solved by) a case base. On the other hand, cases with larger reachability have lower contributions to the competence of a case base, since these cases can be solved (reached) by many other existing cases. Therefore, their significance is evident from the viewpoint of selecting cases for constructing and/or maintaining case bases.

5.3.2 Quantitative Maintenance

Some typical tasks that are to be assured through quantitative maintenance are as follows:

1. *Controlling the size of the case base.* In many CBR applications, the case base grows at a fast rate, and this causes inefficiency in the case retrieval process. Purely adding more cases for problem solving does not guarantee that the solution will be better. On the other hand, deleting too many cases may reduce the competence of the CBR system. Therefore, there is always a trade-off between the size of a case base and its competence. This means retaining the minimum number of cases such that the competence of the system is preserved.

2. *Revising the case indexing structure.* Case representation and indexing structures need to be reviewed and updated periodically due to the changes in domain knowledge (e.g., a different description of the cases) or classifying rules (e.g., the

classifying rules are replaced by new ones). Furthermore, indexes of the new cases may need to be created, and the feature weights for classification may need to be relearned.

3. *Collecting performance statistics.* For scheduling and identifying effective maintenance activities, various performance indicators need to be collected, such as the use of cases, cost of retrieval, and case access frequency. This information is useful in fine tuning of the performance of the CBR system. For example, cases with a high frequency of access will be stored in fast disks or on client nodes over a distributed network.

4. *Detecting irrelevant or obsolete cases.* These are the cases that no longer have value to the system. The solutions of these cases are no longer valid (e.g., obsolete machine parts/components that are no longer manufactured by the supplier).

5. *Cleaning noisy or missing data.* Cases with noisy or missing data are very common in CBR systems because cases are usually recorded and input by different persons and from different sources. The same problem may be described in many different ways, or different problems may be described similarly by different persons. Filling in the missing information and standardizing the case description therefore become important maintenance tasks of these systems.

6. *Gathering user feedback and comments.* The main advantage of CBR systems is that it does not explicitly need the underlying domain theory. The problem-solving ability is based entirely on the similarity assumption (i.e., similar problems should have similar solutions). However, this assumption holds only when feature similarity directly reflects the characteristics of the underlying domain theory. In many practical real-world problems, this similarity assumption needs to be evaluated and reviewed regularly by users. The feedback and comments from them are therefore crucial for CBR developers to rethink and redesign the reasoning strategy behind CBR systems. For example, a new feature weight training scheme may be needed, a new adaptation algorithm may be required, and a new interpretation of similarity may be desirable to make the system up to date.

7. *Performing backup and recovery.* Similar to other information systems, many housekeeping tasks are required for CBR systems, such as backup and recovery of the case base, creation of transaction logs, archiving, access controls, creation of case-base views (i.e., access to certain cases/or fields only), and possibly many others that could be defined by the user.

5.4   CASE-BASE MAINTENANCE USING A ROUGH-FUZZY APPROACH

Recently, with the rapid growth of applying intelligent systems on the World Wide Web (WWW), practitioners are becoming more interested to consider the potential of distributed case-based reasoning (CBR) systems [41–45]. For example, consider a distributed CBR help-desk application over the WWW in which the representative cases need to be stored physically in the client nodes (i.e., for quick access), and all the original cases reside in the central server. To keep this distributed CBR system up to date, a strategy is needed to maintain the cases as well as the reasoning technique (e.g., similarity function and the case-dispatching strategy) used.

For real-world problems, there is always a limit on the size of the client-side case bases, a limit that is based on space and performance trade-off. The question then arises: How should cases be selected for storing on the client side? Usually, the client-side case base will act as a cache (i.e., storing cases that are most likely to be needed by individual users). A simple solution is to store those cases that are similar to such problems that the user has submitted in the past. However, this would guarantee optimal solutions for only a small portion of the queries. A better strategy is needed to select the cases for constructing the client-side case base such that it is competence-rich (i.e., provides comprehensive coverage of the problem space). In the following section we describe a methodology demonstrating how such cases can be selected and updated regularly for building a client-side case base.

5.4.1   Maintaining the Client Case Base

The block diagram of the methodology, consisting of six phases, is shown in Figure 5.2. The aim of phase 1 is to learn the feature weights of the cases. This is done using an evaluation function that does not need to know the class information. Using the information of these weights, the original feature space is transformed into a weighted feature space such that clustering of cases can be done more effectively (i.e., similarity/dissimilarity among cases becomes more apparent). Phase 2 aims to partition the case base into several clusters using the weighted distance metric based on the concepts of intracluster and intercluster similarity. Several representative cases are identified in each of these clusters, and a non-representative case in the cluster is approximated by the representative cases. The approximation is carried out using a group of fuzzy adaptation rules. Phase 3 aims to mine these adaptation rules using a rough-fuzzy method. Rough sets are used for removing the redundant attributes through reducts, thereby reducing the searching time. Fuzzy sets are used to describe the approximation ability of the fuzzy adaptation rules, and those rules having strong approximation ability are then selected as the representative ones, thereby reducing their number. These representative rules are considered as the adaptation knowledge associated with the representative cases. Based on these adaptation rules, phase 4 uses a reasoning mechanism to predict the adjustment of the solution for a query case. Phase 5 describes how to select several representative cases from each cluster based on the concepts of coverage and reachability. Since the sets representing the coverage and reachability of different cases are usually ill-defined, the notion of fuzzy sets has been incorporated. The representative fuzzy adaptation rules and the cases thus obtained then constitute the thin-client-side case base in the distributed CBR system. (Note that the competence of this case base will be similar to that of the

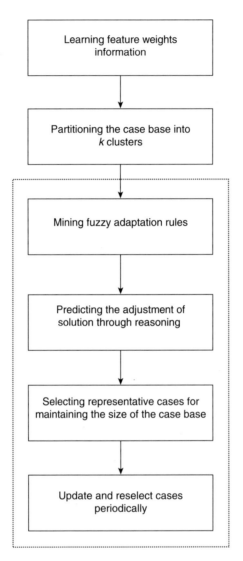

Figure 5.2   Methodology for building and maintaining a client-size case base.

original central server, as discussed earlier.) Phase 6 involves periodical updating and reselection of representative cases because of the changes in domain knowledge and task environment. These are described below with implementation on a real-life data [46]. Here phases 3 to 6 (dashed block in Fig. 5.2) are given the main emphasis, since they are relatively more concerned with the task of case-base maintenance.

5.4.1.1 Learning Feature Weights As described in Section 3.5, there are several supervised and unsupervised methods for learning feature weights. They mainly involve defining an evaluation function such that the smaller the value of the evaluation function, the better the corresponding feature weights. Thus when this function attains its minimum, the best set of feature weights is obtained.

5.4.1.2 Partitioning the Case Base into Several Clusters For a large case base, it is usually necessary to divide it first into several clusters, so that the representative cases can be selected for description of the concepts conveyed by these clusters. In this phase the case base is partitioned into several clusters using the weighted distance metric with the weights learned in phase 1. Since the features considered here are real-valued, any clustering method such as the c-means algorithm, Kohonen's self-organized network [47], or a similarity matrix-based approach [48] can be used.

5.4.1.3 Mining Adaptation Rules by a Rough-Fuzzy Approach Let there be M clusters in the case base obtained in phase 2. For a cluster $LE = \{e_1, e_2, \ldots, e_m\}$, let each case be denoted as $e_i = (x_{i1}, x_{i2}, \ldots, x_{in}, \theta_i)$ for $i = 1, 2, \ldots, m$, where the jth attribute x_{ij} denotes the value of feature F_j $(1 \leq j \leq n)$ of the ith case, and θ_i corresponds to the action or solution (i.e., class label) of the ith case. Compute, for every case e_k in the cluster LE,

$$\phi_{ik} = e_i - e_k = (x_{i1} - x_{k1}, x_{i2} - x_{k2}, \ldots, x_{in} - x_{kn}, \theta_i - \theta_k)$$
$$= \{y_{ik,1}, y_{ik,2}, \ldots, y_{ik,n}, u_{ik}\}, \quad i, k = 1, 2, \ldots, m \quad (5.1)$$

where ϕ_{ik} is called the ith discrepancy vector of the kth case, and all such discrepancy vectors ϕ_{ik} $(i, k = 1, 2, \ldots, m)$ constitute a set denoted by $\{\phi_{ik} | \phi_{ik} \in R^{n+1}, i, k = 1, 2, \ldots, m\}$. $y_{ik,j} = x_{ij} - x_{kj}, j = 1, 2, \ldots, n$, denotes the value of the jth feature-difference attribute of ϕ_{ik}, and $u_{ik} = \theta_i - \theta_k$ is the value of the action difference. Each discrepancy vector ϕ_{ik} is seen to be described by $(n + 1)$ difference attributes $\{\text{Attr}^{(1)}, \text{Attr}^{(2)}, \ldots, \text{Attr}^{(n)}, \text{Attr}^{(n+1)}\}$, where $\text{Attr}^{(n+1)}$ is the solution (action)-difference attribute.

Now we determine a set of membership functions representing fuzzy sets negative big (NB), negative small (NS), zero (ZE), positive small (PS), and positive big (PB) for each difference attribute, construct a set of fuzzy rules corresponding to each case, reduce the number of attributes of these rules using rough sets, and then discard the redundant rules. The resulting rules, thus mined, constitute the adaptation knowledge of the case base.

Constructing Membership Functions Let each of the $(n + 1)$ difference attributes of the discrepancy vector ϕ_{ik} be fuzzified into five linguistic terms: NB, NS, ZE, PS,

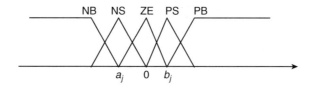

Figure 5.3   Five membership functions.

and PB. Their membership (triangular) functions—$\mu_{NB}, \mu_{NS}, \mu_{ZE}, \mu_{PS}$, and μ_{PB}—along the jth attribute (see Fig. 5.3) are defined by

$$
\mu_{NB}(x) = \begin{cases} 1, & x \leq 2a_j \\ \frac{x}{a_j} - 1, & 2a_j < x < a_j \\ 0, & x \geq a_j \end{cases}
$$

$$
\mu_{NS}(x) = \begin{cases} 0, & x \leq 2a_j, x \geq 0 \\ 2 - \frac{x}{a_j}, & 2a_j < x < a_j \\ \frac{x}{a_j}, & a_j \leq x < 0 \end{cases}
$$

$$
\mu_{ZE}(x) = \begin{cases} 0, & x \leq a_j, x \geq b_j \\ 1 - \frac{x}{a_j}, & a_j < x < 0 \\ 1 - \frac{x}{b_j}, & 0 \leq x < b_j \end{cases} \tag{5.2}
$$

$$
\mu_{PS}(x) = \begin{cases} 0, & x \leq 0, x \geq 2b_j \\ \frac{x}{b_j}, & 0 < x < b_j \\ 2 - \frac{x}{b_j}, & b_j \leq x < 2b_j \end{cases}
$$

$$
\mu_{PB}(x) = \begin{cases} 0, & x \leq b_j \\ \frac{x}{b_j} - 1, & b_j < x < 2b_j \\ 1, & x \geq 2b_j \end{cases}
$$

where

$$
a_j = \frac{\sum_{y \in N_1} y}{\text{Card}(N_1)}
$$

$$
b_j = \frac{\sum_{y \in N_2} y}{\text{Card}(N_2)} \tag{5.3}
$$

and

$$
N_1 = \{y | y \in \text{Range } (\text{Attr}^{(j)}), y < 0\}
$$
$$
N_2 = \text{Range } (\text{Attr}^{(j)}) - N_1
$$
$$
\text{Range } (\text{Attr}^{(j)}) = \{y_{1k,j}, y_{2k,j}, \ldots, y_{(m-1)k,j}\}
$$

and Card($\cdot$) denotes the cardinality of a crisp set ($\cdot$).

After the process of fuzzification of the $(n + 1)$ difference attributes of the discrepancy vector ϕ_{ik} using the membership functions [equations (5.2)], the corresponding fuzzy discrepancy vector φ_{ik} having $5(n + 1)$ components is generated as follows:

$$\varphi_{ik} = \{\mu_{NB}(y_{ik,1}), \mu_{NS}(y_{ik,1}), \mu_{ZE}(y_{ik,1}), \mu_{PS}(y_{ik,1}), \mu_{PB}(y_{ik,1});$$
$$\cdots \mu_{NB}(y_{ik,2}), \ldots, \mu_{PB}(y_{ik,2}); \ldots,$$
$$\mu_{NB}(y_{ik,n}), \ldots, \mu_{PB}(y_{ik,n}); \mu_{NB}(u_{ik}), \ldots, \mu_{PB}(u_{ik})\} \qquad (5.4)$$

where $\mu_{NB}(\cdot), \mu_{NS}(\cdot), \mu_{ZE}(\cdot), \mu_{PS}(\cdot)$, and $\mu_{PB}(\cdot)$ denote the membership values of $(\cdot)$ to the linguistic fuzzy sets NB, NS, ZE, PS, and PB, respectively, and $(\cdot)$ represents either the feature-difference attribute (y) or the solution-difference attribute (u).

Determining Fuzzy Rules (Knowledge Base)   For obtaining the fuzzy rule corresponding to this fuzzy discrepancy vector, we determine its n different antecedent parts (from the n feature-difference attributes) and one consequent part (from the solution-difference attribute) by selecting the respective fuzzy sets corresponding to maximum membership values. For example, if

$$\mu_{NS}(y_{ik,j}) = \max\{\mu_{NB}(y_{ik,j}), \mu_{NS}(y_{ik,j}), \ldots, \mu_{PB}(y_{ik,j})\} \qquad (5.5)$$

then the fuzzy set (i.e., linguistic term) NS is chosen as the label for the jth, $j = 1, 2, \ldots, n$, antecedent of the rule. The case for the consequent part is similar. Therefore, for every case in a cluster $LE = \{e_1, e_2, \ldots, e_m\}$ consisting of m cases, we obtain m fuzzy rules. Let the ith rule, $i = 1, 2, \ldots, m$, corresponding to such a case in a cluster be described in terms of its antecedents and consequent as

$$r_i : \quad \text{IF } [\text{Attr}^{(1)} = A_{i1}] \text{ AND } [\text{Attr}^{(2)} = A_{i2}] \cdots \text{ AND}$$
$$[\text{Attr}^{(n)} = A_{in}] \text{ THEN } [\text{Attr}^{(n+1)} = C_j] \qquad (5.6)$$

where $\text{Attr}^{(1)}, \text{Attr}^{(2)}, \ldots, \text{Attr}^{(n)}$ represent the first, second, $\ldots$, nth feature-difference attributes, respectively. $\text{Attr}^{(n+1)}$ is the solution-difference attribute. $A_{i1}, A_{i2}, \ldots, A_{in}$ and C_j denote different fuzzy labels (sets) representing the antecedents and consequent for the ith rule.

Let the m rules obtained corresponding to a case in LE be as shown in Table 5.1. Here the ith row of the table represents the fuzzy rule $r_i, i = 1, 2, \ldots, m$, taking the form

$$\bigcap_{n}^{p=1} A_{ip} \Rightarrow C_i \qquad (5.7)$$

TABLE 5.1 Fuzzy Knowledge Base

No.	Attr$^{(1)}$	Attr$^{(2)}$	$\cdots$	Attr$^{(n)}$	Attr$^{(n+1)}$	True Degree	Inconsistent Degree
r_1	A_{11}	A_{12}	$\cdots$	A_{1n}	C_1	α_1	β_1
r_2	A_{21}	A_{22}	$\cdots$	A_{2n}	C_2	α_2	β_2
$\vdots$	$\vdots$	$\vdots$	$\vdots$	$\vdots$	$\vdots$	$\vdots$	$\vdots$
r_m	A_{m1}	A_{m2}	$\cdots$	A_{mn}	C_m	α_m	β_m

with true degree α_i (see Definition 5.1) and inconsistent degree β_i (see Definition 5.2). These fuzzy rules corresponding to a case that constitutes what is called its initial fuzzy knowledge base.

Reducing Feature-Difference Attributes of Fuzzy Rules After obtaining the fuzzy knowledge base for each case in LE, we reduce the number of feature-difference attributes of each fuzzy rule through generating its minimal reduct using rough set theory. The reduct of a fuzzy rule r_i (Table 5.1) is obtained by selecting a subset of its antecedents, (e.g., $S = \{A_{ij_1}, A_{ij_2}, \ldots, A_{ij_s}\}$, where $S \subset \{A_{i1}, A_{i2}, \ldots, A_{in}\}$ and $\{j_1, j_2, \ldots, j_s\} \subset \{1, 2, \ldots, n\}$) as the set of condition attributes such that:

1. The modified (reduced) rule r_i is denoted as $\cap_{t=1}^{s} A_{ij_t} \Rightarrow C_i$ (or, $\text{Attr}|_s^i \Rightarrow C_i[\alpha_i, \beta_i]$, for short) with a true degree α_i (Definition 5.1) and an inconsistent degree β_i (Definition 5.2).
2. The cardinality of the subset S is approximately minimal. (Note that the selection of the global minimal is an NP-hard problem.)

Reducing the Number of Fuzzy Rules Here we use the algorithm of Wang and Hong [49] for selecting (mining) a subset of the initial fuzzy rules (Table 5.1) so that they can be used for adaptation. The algorithm performs three tasks. First, it generates a minimal reduct for each fuzzy rule $r_i, i = 1, 2, \ldots, m$. Then, among these minimal reducts (fuzzy rules), a subset of rules corresponding to each ϕ_{ik} is identified such that it can cover the respective discrepancy vector $\phi_{i_k}, i, k = 1, 2, \ldots, m$ (Definition 5.7). Finally, a minimal subset of fuzzy rules is identified among those obtained before, based on the ability of each rule to cover all the fuzzy discrepancy vectors. For example, if rule 1 covers ϕ_1, ϕ_2, and ϕ_5, while rule 2 covers ϕ_2 only, rule 1 will be selected.

Before illustrating the algorithms, we provide seven definitions.

Definition 5.1 [50]. The *true degree* of a fuzzy rule $A \Rightarrow B$ is a measure of the strength of the IF–THEN implication of the rule. The higher the value of α, the stronger the IF–THEN implication. It is defined as $\alpha = \sum_{u \in U} \min(\mu_A(u), \mu_B(u))/\sum_{u \in U} \mu_A(u)$, where A and B are two fuzzy sets.

Definition 5.2 [49]. The *inconsistent degree* of a given fuzzy rule is a measure of the number of possible consequences or actions that this rule could imply. For example, if $A \Rightarrow B, A \Rightarrow C$, and $A \Rightarrow D$, the inconsistent degree of the fuzzy rule $A \Rightarrow B$ is 2. It could be defined formally as $|B|$, where $B = \{j | \text{Attr}|_s^i = \text{Attr}|_s^j, C_i \neq C_j\}$; $|B|$ denotes the number of elements of the set B.

Definition 5.3 [49]. For a given fuzzy rule $\text{Attr}|_s^i \Rightarrow C_i[\alpha_i, \beta_i]$, an attribute $A(A \in S)$ is said to be *dispensable* in the fuzzy rule if $\text{Attr}|_{S-\{A\}}^i \Rightarrow C_i$ has a true degree greater than or equal to σ (a given threshold) and an inconsistent degree less than or equal to β_i . Otherwise, attribute A is indispensable in the rule.

Definition 5.4 [49]. For a given fuzzy rule $\text{Attr}|_s^i \Rightarrow C_i[\alpha_i, \beta_i]$, if all the attributes in S are indispensable, this rule is called *independent*.

Definition 5.5 [49]. A subset of attributes $R(R \subset S)$ is called a *reduct* of the rule $\text{Attr}|_S^i \Rightarrow C_i$ if $\text{Attr}|_R^i \Rightarrow C_i$ is independent. The set of attributes K, which are indispensable in the initial rule $\text{Attr}|_S^i \Rightarrow C_i$, is called the *core* of the initial fuzzy rule.

Definition 5.6 [49]. A *minimal reduct* of an initial fuzzy rule is denoted as $\text{Attr}|_R^i \Rightarrow C_i$, where R is the minimal set of attributes. R should satisfy the property that there does not exist S such that S is a reduct of the initial fuzzy rule with $S \subset R$ and $S \neq R$.

Definition 5.7 [49]. A fuzzy rule $\text{Attr}|_S^i \Rightarrow C_i \left[\alpha_i, \beta_i\right]$ is said to *cover* a discrepancy vector if the membership degree of attributes and the membership degree of classification for the vector are all greater than or equal to η (a threshold).

Based on these definitions, the algorithm [49], having three tasks, is described as follows:

Task 1. Reduce the number of attributes in the initial fuzzy rules. This has six steps.

Step 1. For the ith initial fuzzy rule $(1 \leq i \leq m)$, the core K can be determined by verifying whether or not an attribute is dispensable in the attribute set. (Core K can be empty.) Set $\Gamma = 1$.

Step 2. Take Γ attributes $\text{Attr}_1, \text{Attr}_2, \ldots, \text{Attr}_\Gamma$ from $A^i - K$, where $A^i = \{A_{i1}, A_{i2}, \ldots, A_{in}\}$.

Step 3. Add the attributes $\text{Attr}_1, \text{Attr}_2, \ldots, \text{Attr}_\Gamma$ to K.

Step 4. Compute the true degree and the inconsistent degree of the fuzzy rule $\text{Attr}|_K^i \Rightarrow C_i$.

Step 5. If K is a reduction, exit successfully; else some new Γ attributes $\text{Attr}_1, \text{Attr}_2, \ldots, \text{Attr}_\Gamma$ are taken from $A^i - K$, and go to step 3.

Step 6. If all the combinations of elements in $A^i - K$ have been used and a reduction is not found, set $\Gamma = \Gamma + 1$ (K returns to its original state), and go to step 2.

Task 2. Identify a set of fuzzy rules that cover each discrepancy vector. For each $i(1 \leq i \leq m)$ and the set of rules R from task 1, where $R = \{r_1, r_2, \ldots, r_m\}$, and r_i is the minimal reduction of the ith initial rule, R_i, a subset of R, can be determined by checking whether the rule covers the discrepancy vector ϕ_{ik}:

$$R_i = \{r_j | r_j \in R, r_j \text{ covers } \phi_{ik}\}, \qquad i, j = 1, 2, \ldots, m \qquad (5.8)$$

Task 3. Select the fuzzy rules as the adaptation rules. Take $\Omega = \{R_1, R_2, \ldots, R_m\}$, where $R_i(i = 1, 2, \ldots, m)$ is defined in task 2. Let the initial value of R^* (used below) be an empty set. Repeat the following three steps until Ω become empty.

Step 1. For each $r_i \in R$ determined by task 1, compute the number of times that r_i appears in the family Ω.

Step 2. Select r^* such that the number of times r^* appears in the family Ω is the maximum.

Step 3. For $i = 1, 2, \ldots, m$, remove R_i from Ω if $r^* \in R_i$, and replace R^* with $\{r^*\} \cup R^*$. Return to step 1 and repeat until Ω becomes empty.

The algorithm thus generates, through pruning of attributes and rules, for all the cases of a cluster, a set of adaptation rules (fuzzy production rules) R^*. These constitute what is called the *final knowledge base* of the concerned cluster. Note that the values of the thresholds σ and η, which control the number of attributes and the rules, are problem dependent. In the next section we describe a reasoning mechanism, based on these adaptation rules, to predict the amount of adjustment required for the solution of a query (unknown case) case.

5.4.1.4 Predicting Adjustment through Reasoning and Adaptation

Suppose that for a particular case $e_k = (x_{k1}, x_{k2}, \ldots, x_{kn}, \theta_k)$, l^* fuzzy adaptation rules have been extracted, denoted by $\{r_i, i = 1, 2, \ldots, l^*\}$. Let the ith fuzzy adaptation rule $r_i(1 \leq i \leq l^*)$ be represented in the following form:

$$\text{IF } [\text{Atrr}^{(1)} = A_{i1}] \text{ AND } [\text{Atrr}^{(2)} = A_{i2}] \text{ AND} \cdots \text{AND}$$
$$[\text{Atrr}^{(n)} = A_{in}] \text{ THEN } [\text{Attr}^{(n+1)} = C_i] \qquad (5.9)$$

in which $\text{Attr}^{(j)}(j = 1, 2, \ldots, n+1)$, $A_{ij}(j = 1, 2, \ldots, n)$, and C_i are as in equation (5.6). Note that some of the antecedents could be absent because of elimination process stated before.

Let $e_q = \{x_{q1}, x_{q2}, \ldots, x_{qn}, \theta_q\}$ be an input query case, where the attribute values x_j ($1 \leq j \leq n$) are known, whereas the action (solution) θ_q is unknown and needs to be decided. Assuming that the case $e_k = (x_{k1}, x_{k2}, \ldots, x_{kn}, \theta_k)$ has been selected to provide the solution (i.e., to predict the value of unknown θ_q) of the query case e_q, we compute the discrepancy vector ϕ between cases e and e_k, as in equation (5.1):

$$\phi_{qk} = e_q - e_k = (x_{q1} - x_{k1}, x_{q2} - x_{k2}, \ldots, x_{qn} - x_{kn}, \theta_q - \theta_k)$$
$$= \{y_{qk,1}, y_{qk,2}, \ldots, y_{qk,n}, u_{qk}\} \qquad (5.10)$$

Here the feature-difference values $y_{qk,j}(1 \leq j \leq n)$ are known, while the solution-difference value u_{qk} is unknown. The value of u_{qk} is determined by the following procedure:

Step 1. For each fuzzy adaptation rule $r_i(1 \leq i \leq l^*)$ associated with e_k, compute the membership values (i.e., μ_{ij} values) of $y_{qk,j} \in A_{ij}(1 \leq j \leq n)$ of the fuzzy sets corresponding to antecedents $[\mathrm{Attr}^{(j)} = A_{ij}]$ of r_i, with $A_{ij} \neq \emptyset$ ($\emptyset$ is an empty set, $1 \leq j \leq n$).

Step 2. Compute the overall degree of similarity, $\mathrm{SM}^{(i)}$, of the feature-difference vector $\{y_{qk,j}\}$ with respect to the antecedents of the ith rule r_i as

$$\mathrm{SM}^{(i)} = \min_j(\mu_{ij}), \qquad i = 1, 2, \dots, l^*; \quad j = 1, 2, \dots, n \qquad (5.11)$$

Step 3. Compute the overall degree of similarity, SM_l, of the feature-difference vector $\{y_{qk,j}\}$ with respect to the antecedents of all the rules having the same consequent fuzzy set C_l $(l = 1, 2, \dots, 5)$ as

$$\mathrm{SM}_l = \max_i\{\mathrm{SM}^{(i)} | \text{consequent fuzzy set} = C_l\} \qquad (5.12)$$

where C_l $(l = 1, 2, \dots, 5)$ represents the five fuzzy sets NB, NS, ZE, PS, and PB (Fig. 5.3).

Step 4. Compute the solution-difference u_{qk} [equation (5.10)] using the defuzzification formula

$$u_{qk} = \frac{2a_{n+1}\mathrm{SM}_1 + a_{n+1}\mathrm{SM}_2 + b_{n+1}\mathrm{SM}_4 + 2b_{n+1}\mathrm{SM}_5}{\mathrm{SM}_1 + \mathrm{SM}_2 + \cdots + \mathrm{SM}_5} \qquad (5.13)$$

where the parameters a_{n+1} and b_{n+1}, corresponding to the solution attribute, are defined by equation (5.3).

Step 5. Compute the estimated solution $\hat{\theta}_q$ of case e_q as

$$\hat{\theta}_q = \theta_k + u_{qk} \qquad (5.14)$$

5.4.1.5 *Selecting Representative Cases*   Here we describe a methodology for selecting representative cases from each cluster for the purpose of maintaining the size of a case base using the adaptation rules obtained in the preceding phase. The selection strategy is based on the concepts of ε-*coverage* and ε-*reliability* [39,51]. Let LE be a cluster of m cases where each case e is accompanied by a set of adaptation rules $R(e)$, let ε be a small positive number, and let $e_p = (x_{p1}, x_{p2}, \dots, x_{pn}, \theta_p)$ and $e_q = (x_{q1}, x_{q2}, \dots, x_{pn}, \theta_q)$ be two cases in the cluster LE. Then e_p is said to ε-*cover* e_q, or e_q is said to be ε-*covered* by e_p if

$$|\theta_p + u_{qp} - \theta_q| = |\hat{\theta}_q - \theta_q| \leq \varepsilon \qquad (5.15)$$

where u_{qp}, the solution difference obtained by equation (5.13) using the adaptation rules associated with e_p, is used to estimate the solution of e_q (i.e., $\hat{\theta}_q$). The contribution of a case e_p (with its associated fuzzy rules) to the competence of (i.e., the range of problems that can be solved by) the case base can be characterized by the ε-CoverageSet and the ε-ReachabilitySet of e_p. These are defined as

$$\varepsilon\text{-CoverageSet}(e_p) \quad = \{e|e \in \text{LE}, e \text{ is } \varepsilon\text{-covered by } e_p\} \qquad (5.16)$$
$$\varepsilon\text{-ReachabilitySet}(e_p) = \{e|e \in \text{LE}, e \, \varepsilon\text{-covers } e_p\} \qquad (5.17)$$

The ε-CoverageSet(e_p) therefore represents the generalization capability of the case e_p. As the number of cases in the ε-CoverageSet(e_p) increases, the significance of e_p in representing the cluster increases. On the other hand, the ε-ReachabilitySet(e_p) represents the degree to which the case e_p can be replaced by other case(s). Therefore, the smaller the number of cases in the ε-ReachabilitySet(e_p), the higher the significance of e_p in representing the cluster.

Based on these measures, we explain below an algorithm for eliminating some of the cases in LE, thereby retaining only the significant ones as the representative cases for the purpose of maintaining the size of a case base.

Algorithm. Given the cluster LE and an ε value, let B be the set of the representative cases of LE and be initialized to an empty set.

Step 1. For each case e in LE, determine ε-CoverageSet(e) [equation (5.16)] and ε-ReachabilitySet(e) [equation (5.17)] by a set of adaptation rules $R(e)$ associated with the case e.

Step 2. Find the case(s) e^* such that

$$|\varepsilon\text{-CoverageSet}(e^*)| = \max {}_e|\varepsilon\text{-CoverageSet}(e)|, \qquad e, e^* \in \text{LE}$$

Let E^* be the set of such case(s) e^*. If $|E^*| = 1$, go to step 3; otherwise, select the case(s) e^{**} from E^* such that

$$|\varepsilon\text{-ReachabilitySet}(e^{**})| = \min {}_e|\varepsilon\text{-ReachabilitySet}(e^*)|, \qquad e^{**}, e^* \in \text{LE}$$

Let E^{**} be the set of such case(s) e^{**}. If $|E^{**}| = 1$, set case $e^* = e^{**}$; otherwise, select one of the cases randomly from e^{**} as case e^*.

Step 3. Set $B = B \cup \{e^*\}$ and $\text{LE} = \text{LE} - \varepsilon\text{-CoverageSet } (e^*)$. If $|LE| = 0$, stop; else, set $E^* = \emptyset$ and $E^{**} = \emptyset$; then go to step 2.

B provides the final set of representative cases for the cluster LE.

5.4.1.6 *Updating and Reselecting Cases*   After the case base is constructed through the selection of representative cases, it is required to maintain it through periodic reselection and updating of cases, because of changes in the environment/

domain, by repeatedly applying the tasks described in the preceding five phases. For example, consider the thin-client case base (Fig. 5.2) in a distributed help-desk application. Due to various changes in user requirements, one needs to keep the case base up to date through reselecting new representative cases. Note that the value of ε controls the number of representative cases in a cluster, and therefore controls the size of the case base. The size of the case base increases as ε increases.

5.4.1.7 *Examples*
Here we demonstrate through computation on examples the various steps of the aforesaid algorithms for mining adaptation rules, obtaining the solution of a query case by applying them, and then selecting representative cases. As a first example, let $LE = \{e_1, e_2, e_3, e_4, e_5\}$ be one of the clusters, and each case be represented by three features and an action (solution), such as $e_1 = (3,-1,4,2)$, $e_2 = (-2,3,1,-1)$, $e_3 = (4,7,2,3)$, $e_4 = (-3,-3,9,6)$, and $e_5 = (5,-6,6,1)$.

Constructing Membership Functions   Let us consider first the case e_1; compute the difference between it and the other cases in LE. This generates five discrepancy vectors (each having three feature-difference attributes, $\text{Attr}^{(1)}, \text{Attr}^{(2)}, \text{Attr}^{(3)}$, and one solution-difference attribute, $\text{Attr}^{(4)}$) as

$$\begin{aligned}
\phi_{11} &= e_1 - e_1 = (0,0,0,0), & \phi_{21} &= e_2 - e_1 = (-5,4,-3,-3), \\
\phi_{31} &= e_3 - e_1 = (1,8,-2,1), & \phi_{41} &= e_4 - e_1 = (-6,-2,5,4), \\
\phi_{51} &= e_5 - e_1 = (2,-5,2,-1)
\end{aligned}$$

Here the ranges of these difference attributes are as follows:

$$\text{Range}(\text{Attr}^{(1)}) = \{-6,-5,0,1,2\}$$
$$\text{Range}(\text{Attr}^{(2)}) = \{-5,-2,0,4,8\}$$
$$\text{Range}(\text{Attr}^{(3)}) = \{-3,-2,0,2,5\}$$
$$\text{Range}(\text{Attr}^{(4)}) = \{-3,-1,0,1,4\}$$

Then for each attribute $\text{Attr}^{(j)}$ ($j = 1,2,3,4$), five membership functions, corresponding to fuzzy sets NB, NS, ZE, PS, and PB (as shown in Figs. 5.4–5.7), are determined by computing the parameters a_j and b_j ($j = 1,2,3,4$) as

$$\text{Attr}^{(1)}: \quad a_1 = \frac{-6-5}{2} = -5.5, \quad b_1 = \frac{0+1+2}{3} = 1$$

$$\text{Attr}^{(2)}: \quad a_2 = \frac{-5-2}{2} = -3.5, \quad b_2 = \frac{0+4+8}{3} = 4$$

$$\text{Attr}^{(3)}: \quad a_3 = \frac{-3-2}{2} = -2.5, \quad b_3 = \frac{0+2+5}{3} = 2.3$$

$$\text{Attr}^{(4)}: \quad a_4 = \frac{-3-1}{2} = -2, \quad b_4 = \frac{0+1+4}{3} = 1.7$$

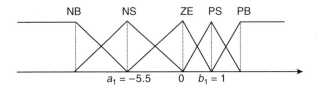

Figure 5.4   Membership function along Attr$^{(1)}$.

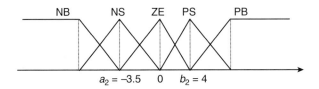

Figure 5.5   Membership function along Attr$^{(2)}$.

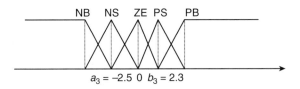

Figure 5.6   Membership function along Attr$^{(3)}$.

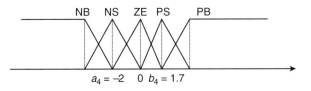

Figure 5.7   Membership function along Attr$^{(4)}$.

Then the membership values for each component of these discrepancy vectors corresponding to these five fuzzy sets are computed. Consider ϕ_{21}, for example. There are five membership values for each attribute Attr$^{(j)}$ ($j = 1,2,3,4$), and they are listed in Table 5.2.

Determining Fuzzy Rules   Consider the first row in Table 5.2. According to equation (5.5), determine the fuzzy set with maximum membership value, NS, as the fuzzy label of the first antecedent in r_2. Similarly, PS, NS, and NS are selected

TABLE 5.2 Membership Values of ϕ_{21}

ϕ_{21}	NB	NS	ZE	PS	PB
$Attr^{(1)}$	0	0.9	0.1	0	0
$Attr^{(2)}$	0	0	0	1	0
$Attr^{(3)}$	0.2	0.8	0	0	0
$Attr^{(4)}$	0.5	0.5	0	0	0

as the labels for other two antecedents and the consequent of r_2, respectively, therefore, we write

$$r_2 : \quad \text{IF } [Attr^{(1)} = NS] \text{ AND } [Attr^{(2)} = PS] \text{ AND } [Attr^{(3)} = NS] \text{ THEN } [Attr^{(4)} = NS]$$

Similarly, the fuzzy rules based on the other three discrepancy vectors are generated as follows:

$$r_3 : \quad \text{IF } [Attr^{(1)} = PS] \text{ AND } [Attr^{(2)} = PB] \text{ AND } [Attr^{(3)} = NS] \text{ THEN } [Attr^{(4)} = PS]$$

$$r_4 : \quad \text{IF } [Attr^{(1)} = NS] \text{ AND } [Attr^{(2)} = NS] \text{ AND } [Attr^{(3)} = PB] \text{ THEN } [Attr^{(4)} = PB]$$

$$r_5 : \quad \text{IF } [Attr^{(1)} = PB] \text{ AND } [Attr^{(2)} = NS] \text{ AND } [Attr^{(3)} = PS] \text{ THEN } [Attr^{(4)} = ZE(NS)]$$

Note that in ϕ_{21}, since the membership values of $Attr^{(4)}$ to NB and NS are found to be equal ($= 0.5$), we randomly choose any of them, say NS, as the label for the consequent. Similarly, ZE is selected as the consequent part of r_5. Moreover, the fuzzy rule r_1 corresponding to ϕ_{11}, as expected, is

$$\text{IF } [Attr^{(1)} = Attr^{(2)} = Attr^{(3)} = ZE] \text{ THEN } [Attr^{(4)} = ZE]$$

Reducing Feature-Difference Attributes of Fuzzy Rules Consider r_2 as an example. Its true degree and inconsistent degree are found, respectively, as $\alpha = 0.63$ and $\beta = 0$. Assuming that the threshold of the true degree $\sigma = 0.62$, it is found that by removing $Attr^{(1)} = NS$ or/and $Attr^{(2)} = PS$, we still have the same values for true degree and inconsistent degree: $\alpha' = 0.63$ and $\beta' = 0$. Since $\alpha' > \sigma$, and $\beta' = \beta$, $Attr^{(1)}$ and $Attr^{(2)}$ are dispensable attributes, but if they are removed at the same time, we have a low value, 0.31, for the true degree. In this situation, only one of them, say $Attr^{(1)}$ (which is chosen randomly), can be removed. On the other hand, if we remove $Attr^{(3)} = NS$, the true degree and inconsistent degree are 0.56 and 0, respectively. Since $0.56 < \sigma$, $Attr^{(3)}$ is considered to be an indispensable attribute for rule r_2. Removing $Attr^{(1)}$ from r_2, we have the modified rule, in terms of the following reduct, as

$$r_2 : \quad \text{IF } [Attr^{(2)} = PS] \text{ AND } [Attr^{(3)} = NS] \text{ THEN } [Attr^{(4)} = NS]$$

Similarly, the reducts of the other modified rules, together with the same σ value, are as follows:

r_3 :   IF $[\text{Attr}^{(1)} = \text{PS}]$ AND $[\text{Attr}^{(3)} = \text{NS}]$ THEN $[\text{Attr}^{(4)} = \text{PS}]$

r_4 :   IF $[\text{Attr}^{(1)} = \text{NS}]$ AND $[\text{Attr}^{(3)} = \text{PB}]$ THEN $[\text{Attr}^{(4)} = \text{PB}]$

r_5 :   IF $[\text{Attr}^{(2)} = \text{NS}]$ AND $[\text{Attr}^{(3)} = \text{PS}]$ THEN $[\text{Attr}^{(4)} = \text{ZE}]$

Reducing the Number of Fuzzy Rules   Consider ϕ_{21} as an example. Let the threshold be $\eta = 0.5$. For the rule r_2 : IF $[\text{Attr}^{(2)} = \text{PS}]$ AND $[\text{Attr}^{(3)} = \text{NS}]$ THEN $[\text{Attr}^{(4)} = \text{NS}]$, the degree of membership of $\text{Attr}^{(2)}$ to PS is 0.8 and $\text{Attr}^{(3)}$ to NS is 0.5. Since all the membership values are greater than or equal to the threshold $\eta(= 0.5)$, r_2 covers ϕ_{21} [equation (5.8)]. Similarly, one can check if there are other rules that can cover ϕ_{21}. Finally, it is only r_2 that is found to cover ϕ_{21}; that is, $R_2 = \{r_2\}$.

Similarly, we have $R_1 = \{r_1\}, R_3 = \{r_3\}, R_4 = \{r_4\}$, and $R_5 = \{r_5\}$. Here each discrepancy vector is seen to be covered only by its corresponding rule. Therefore, all these rules are selected as the adaptation rules of e_1 for covering all the vectors $\phi_{i1} (i = 1, 2, \ldots, 5)$ and $R^* = \{r_1, r_2, r_3, r_4, r_5\}$.

In the example above, since all the rules are retained, the various steps involved in reducing the number of rules could not be shown. To illustrate these steps, we give another example.

Let us consider a situation where a discrepancy vector is covered only by its corresponding rule or several other rules, such as $R_1 = \{r_1\}, R_2 = \{r_2, r_3\}, R_3 = \{r_2, r_3, r_4\}, R_4 = \{r_4\}$, and $R_5 = \{r_3, r_5\}$. Here the number of occurrences of the rules $r_i (i = 1, 2, \ldots, 5)$ in Ω are 1, 2, 3, 1, and 1, respectively. Since r_3 has occurred maximum and it covers $\phi_{21}(R_2), \phi_{31}(R_3)$, and $\phi_{51}(R_5)$, we first select r_3 and remove R_2, R_3, and R_5 from Ω. Similarly, r_1 and r_4 are then selected, and R_1 and R_4 are removed from Ω. Since Ω becomes empty, the set of adaptation rules is $R^* = \{r_1, r_3, r_4\}$.

Predicting Adjustment through Reasoning   Let $e_q = (4, 1, 3, \theta_q)$ be a query case, where θ_q is the unknown solution, and e_1 be selected to solve the query case. The discrepancy vector ϕ_{q1} is

$$\phi_{q1} = e_q - e_1 = (y_{q1,1}, y_{q1,2}, y_{q1,3}, u_{q1}) = (1, 2, -1, u_{q1}), \qquad \text{where } u_{q1} = \theta_q - 2$$

The solution-difference u_{q1} is determined as follows: Consider r_3(IF$[\text{Attr}^{(1)} = \text{PS}]$ AND $[\text{Attr}^{(3)} = \text{NS}]$ THEN $[\text{Attr}^{(4)} = \text{PS}]$, as described before, where we have two membership values: $1(y_{q1,1} \in \text{PS})$, and $0.5(y_{q1,3} \in \text{NS})$. Then using equation (5.11), we get

$$\text{SM}^{(3)} = \min\{1, 0.5\} = 0.5$$

Similarly, $SM^{(1)} = 0, SM^{(2)} = 0.4, SM^{(4)} = 0$, and $SM^{(5)} = 0$. Compute SM_l ($l = 1, 2, \ldots, 5$) by equation (5.12); we have for $i = 1, 2, \ldots, 5$,

$$SM_1 = \max_i\{SM^{(i)}|Attr^{(4)} = NB\} = 0$$
$$SM_2 = \max_i\{SM^{(i)}|Attr^{(4)} = NS\} = SM^{(2)} = 0.4$$
$$SM_3 = \max_i\{SM^{(i)}|Attr^{(4)} = ZE\} = \max\{SM^{(1)}, SM^{(5)}\} = 0$$
$$SM_4 = \max_i\{SM^{(i)}|Attr^{(4)} = PS\} = SM^{(3)} = 0.5$$
$$SM_5 = \max_i\{SM^{(i)}|Attr^{(4)} = PB\} = SM^{(4)} = 0$$

Note that since the label NB does not occur in the consequents of any of the adaptation rules, SM_1 is assumed to be 0.

Then the solution-difference u_{q1} is computed [equation (5.13)] as

$$u_{q1} = (-2) \times 0.4 + 1.7 \times 0.5 = 0.05$$

Thus, the solution of the query case e_q is [equation (5.14)]

$$\theta_q = 2 + u_{q1} = 2.05$$

Case Selection   Let $LE = \{e_1, e_2, e_3, e_4\}, e_1 = (1,4,3), e_2 = (2,5,2), e_3 = (-1,6, -2), e_4 = (3,-7,-5)$, and $\varepsilon = 0.2$. Consider e_1, for example. The solution differences of e_j, u_{j1} ($j = 2, 3, 4$) are as follows [equation (5.13)]: $u_{21} = 0.5, u_{31} = -5$, and $u_{41} = -4$. Since

$$|\theta_1 + u_{21} - \theta_2| = |\hat{\theta}_2 - \theta_2| = 3 + 0.5 - 2 = 1.5 > \varepsilon(= 0.2)$$
$$|\theta_1 + u_{31} - \theta_3| = |\hat{\theta}_3 - \theta_3| = 3 - 5 + 2 = 0 < \varepsilon$$
$$|\theta_1 + u_{41} - \theta_4| = |\hat{\theta}_4 - \theta_4| = 3 - 4 + 5 = 4 > \varepsilon$$

according to equation (5.16), $e_1\varepsilon$-covers e_3, and e_1 certainly covers itself. That is,

$$\varepsilon\text{-CoverageSet}(e_1) = \{e_1, e_3\}$$

Similarly, the ε-CoverageSets of the other cases are determined as

$$\varepsilon\text{-CoverageSet}(e_2) = \{e_1, e_2\}$$
$$\varepsilon\text{-CoverageSet}(e_3) = \{e_3\}$$
$$\varepsilon\text{-CoverageSet}(e_4) = \{e_4\}$$

TABLE 5.3 Sample Record of the Travel Case Base

Name	Data Type	Example
Holiday type	Symbolic	Recreation
Number of persons	Numeric	6
Region	Symbolic	France
Transportation	Symbolic	Car
Duration	Numeric	14
Season	Symbolic	January
Accommodation	Symbolic	Holiday Flat
Hotel	Symbolic	H. Flat, Cheval Blanc, France
Price	Numeric	1728

Consequently, the ε-ReachabilitySet of each case is determined [equation (5.17)] as

$$\varepsilon\text{-ReachabilitySet}(e_1) = \{e_1, e_2\}$$
$$\varepsilon\text{-ReachabilitySet}(e_2) = \{e_2\}$$
$$\varepsilon\text{-ReachabilitySet}(e_3) = \{e_1, e_3\}$$
$$\varepsilon\text{-ReachabilitySet}(e_4) = \{e_4\}$$

Here e_1 and e_2 are found to have the greatest coverage. Since the number of cases in their ε-CoverageSets is the same, the case with less reachability (i.e., e_2), is selected. Then the cases that can be covered by it (i.e., e_1 and e_2), are removed from LE. Repeat this process, and then select e_4 and e_3; this ends the algorithm. Therefore, the final set of the representative cases for LE is obtained as $\{e_2, e_3, e_4\}$.

5.4.2 Experimental Results

In this section we present some of the experimental results [46] of the case-base maintenance method described in Section 5.4.1 using a set of test cases from the travel domain. The case base Travel has 1024 cases and is available from the Web site *http//www.ai-cbr.org*. Each travel case consists of nine attributes: type of vacation, length of stay, holiday type, hotel, and so on. Table 5.3 shows a sample record with the feature "Price" used as the solution feature.

Feature weights are determined as noted in Section 5.4.1.1. Table 5.4 shows such values when the gradient descent technique was used (with the number of learning iterations $= 10,000$) as an optimization tool. The weighted Travel case base is then

TABLE 5.4 Feature Weights of the Problem Features

Type	No. of Persons	Region	Trans.	Duration	Season	Accom.	Hotel
0.1374	0.0891	0.0662	0.3691	1.0000	0.0440	0.3443	0.0503

TABLE 5.5 Clusters of the Travel Case Base

Cluster	Number of Cases	Odd or Not-Odd Class
1	40	Not-odd
2	10	Not-odd
3	31	Not-odd
4	76	Not-odd
5	53	Not-odd
6	18	Not-odd
7	13	Not-odd
8	2	Odd
9	32	Not-odd
10	69	Not-odd
11	78	Not-odd
12	116	Not-odd
13	228	Not-odd
.	.	.
.	.	.
.	.	.
53	3	Odd
54	1	Odd
55	2	Odd

partitioned (as mentioned in Section 5.4.1.2) into 55 clusters (see Table 5.5) using a similarity matrix–based clustering algorithm. For the convenience of readers, we describe below the algorithm that first transforms the similarity matrix, computed based on the similarity between cases, to an equivalent matrix (i.e., the final similarity matrix SM in step 5 of the following algorithm), and the cases that are equivalent to each other are considered to be in the same cluster.

Clustering Algorithm

Step 1. Define a significant level (threshold) $\beta \in (0, 1]$.

Step 2. Determine the similarity matrix $SM_{N \times N} = (SM_{pq}^{(w)})$ according to equations (3.1) and (3.2).

Step 3. Compute $SM1 = SM \circ SM = (s_{pq})$, where $s_{pq} = \max_k(\min(SM_{pk}^{(w)}, SM_{kq}^{(w)}))$.

Step 4. If $SM1 = SM$, go to step 5; else, replace SM with SM1 and go to step 3.

Step 5. Determine the clusters based on the rule "case p and case q belong to the same cluster if and only if $SM_{pq} \geq \beta$ in the final similarity matrix SM," and compute the number M of clusters.

The following example illustrates the foregoing clustering algorithm. Let us consider three cases, $e_1 = (1,2), e_2 = (1,4),$ and $e_3 = (1,12),$ and assume that their feature weights w and the parameter α are equal to 1. Then according to equations (3.1) and (3.2), the similarity matrix (i.e., in step 2) is

$$SM_{3\times3} = \begin{bmatrix} 1 & \frac{1}{3} & \frac{1}{11} \\ \frac{1}{3} & 1 & \frac{1}{9} \\ \frac{1}{11} & \frac{1}{9} & 1 \end{bmatrix}$$

Since $s_{pq} = \max_k(\min(SM_{pk}^{(w)}, SM_{kq}^{(w)}))$, SM1 is equal to

$$SM1 = \begin{bmatrix} 1 & \frac{1}{3} & \frac{1}{9} \\ \frac{1}{3} & 1 & \frac{1}{9} \\ \frac{1}{9} & \frac{1}{9} & 1 \end{bmatrix}$$

The next step is to compare SM1 with SM. Since the current SM1 $\neq$ SM, replace SM with SM1:

$$SM_{3\times3} = SM1 = \begin{bmatrix} 1 & \frac{1}{3} & \frac{1}{9} \\ \frac{1}{3} & 1 & \frac{1}{9} \\ \frac{1}{9} & \frac{1}{9} & 1 \end{bmatrix}$$

Recalculate SM1 until SM1 = SM, and finally, we get SM1 (i.e., the equivalent matrix) as

$$SM1 = \begin{bmatrix} 1 & \frac{1}{3} & \frac{1}{9} \\ \frac{1}{3} & 1 & \frac{1}{9} \\ \frac{1}{9} & \frac{1}{9} & 1 \end{bmatrix}$$

After obtaining the equivalent matrix above, clusters can be determined according to the threshold β.

If $\beta = \frac{1}{5}$, case e_1 and case e_2 belong to the same cluster, and case e_3 forms an individual cluster, and therefore $M = 2$. On the other hand, if $\beta = \frac{1}{2}$, each case e_1, e_2, and e_3 forms an individual cluster, respectively, and therefore $M = 3$. Since the value of M depends largely on the selection of the value of β, the heuristic rule [equation (3.31)] based on the concepts of intercluster similarity and intracluster similarity is used to determine the best value of M.

Some of these clusters are shown in Table 5.5. Clusters with less than 10 cases are labeled as odd clusters and the others as not-odd clusters. The learning of fuzzy adaptation rules is carried out only on the not-odd clusters. In the process of mining of fuzzy adaptation rules, only the numeric features are fuzzified into five linguistic variables (Fig. 5.3): NB, NS, ZE, PS, and PB.

TABLE 5.6   Reachability and Coverage of Each Case in Cluster 7 of the Travel Case Base

Case	Number of Cases Covered by Case	Actual Cases Covered by Case	Number of Adaptation Rules
0	3	4,7,12	2
1	7	2,3,4,5,6,7,10,12	5
2	4	3,5,6	4
3	4	0,2,4,7	7
4	2	0,6	3
5	2	0,4	4
6	2	2,5	5
7	5	0,2,3,4,5	5
8	6	0,2,3,5,6,12	3
9	6	2,3,4,5,10,12	5
10	1	0	4
11	5	2,3,4,6,12	5
12	5	0,2,3,4,6	7

An example of a fuzzy adaptation rule (with both numeric and symbolic features) generated from the rough-fuzzy method is as follows:

IF $[\text{Attr}^{(1)} = [\text{small}|\text{medium}|\text{big}|\text{symbolic values}]]$

AND $[\text{Attr}^{(2)} = [\text{NB}|\text{NS}|\text{ZE}|\text{PS}|\text{PB}]]$

AND $[\text{Attr}^{(3)} = [\text{small}|\text{medium}|\text{big}|\text{symbolic values}]]$

THEN the change of $\text{Price}(\text{Attr}^{(9)}) = [\text{NB}|\text{NS}|\text{ZE}|\text{PS}|\text{PB}]$

Here case $e = \{$holiday type, number of persons, region, transportation, duration, season, accommodation, hotel, and price$\}$, and $\text{Attr}^{(i)}(i = 1, 2, \ldots, 9)$ is a feature-difference attribute of case e. For simplicity reason, the number of antecedents of each fuzzy rule is limited to one in this experiment. For example, in cluster 7 (see Table 5.5), which consists of 13 cases, one of the adaptation rules (with symbolic antecedents and fuzzy action/ solution) is: IF holiday type is changed from "education" to "city," THEN the change of price is positive small.

According to the case-selecting strategy based on reachability and coverage defined in Section 5.4.1.5 with $\varepsilon = 0.05$, cases $\{1,8,9,11\}$ are selected as the representative cases in cluster 7 (see Table 5.6). It means that of the 13 cases in this cluster, nine cases have been deleted (see Table 5.7) based on the selection strategy. As a result of this process, a total of $18(= 5 + 3 + 5 + 5)$ fuzzy adaptation rules are selected. After applying the case selection strategy to each not-odd cluster, 399 cases are deleted altogether out of 933. In other words, the number of cases in the Travel case base can be reduced by 43%, and this smaller case base could be resided in the client CBR system. A further calculation of the relative errors

TABLE 5.7 **Selection of Representative Cases in All the Not-Odd Clusters**

Cluster	Number of Cases	Number of Representative Cases	Number of Cases Deleted	Average Relative Error of the Cases Deleted (%)
1	40	19	21	6.98
2	10	4	6	1.63
3	31	16	15	4.52
4	76	40	36	5.78
5	53	23	30	5.99
6	18	17	1	0.78
7	13	4	9	3.15
9	32	19	13	3.92
10	69	56	13	4.77
11	78	50	28	6.90
12	116	78	38	6.54
13	228	130	98	8.98
17	30	10	20	6.40
19	12	3	9	6.71
20	12	8	4	1.52
23	24	14	10	3.40
25	41	16	25	5.67
28	26	20	6	2.64
31	13	4	9	3.05
33	11	3	8	3.80
	933	534	399	Overall average 6.315%

[equation (5.18)] in this smaller case base indicates that (see the last column and last row of Table 5.7) the solution generated from it is almost as accurate as that generated from the original case base (i.e., about 94% accuracy). If more tolerance of the accuracy is allowed in the selection strategy, a substantially smaller case base can be generated.

$$\text{Relative error} = \frac{\text{actual value of the solution} - \text{computed value}}{\text{actual value of the solution}} \times 100\% \quad (5.18)$$

5.4.3 Complexity Issues

In the method described in Section 5.4.1, the main idea is to transform a large case base to a smaller case base together with a group of fuzzy adaptation rules for the benefit of selecting a competence-rich case base for the client-side CBR system. To provide a complete picture of this method, we provide an analysis of the time and space complexities of this approach.

The time complexity is determined by the number of multiplication, division, maximize and minimize operations required. In the first phase (i.e., learning feature weights; Section 5.4.1.1), the feature evaluation index function is differentiable (smooth) and the searching technique used is gradient descent; it can be guaranteed that the training algorithm is convergent if the learning rate is appropriately small. The total time complexity of the training algorithm is $O(m^2)$, where m is the number of cases.

In the second phase (i.e., partitioning cases into clusters; Section 5.4.1.2), the clustering algorithm mainly involves multiplication of two similarity matrices. Therefore, the time complexity is equal to $O(m^2)$.

The time complexity of the third phase (i.e., the mining fuzzy rule using the rough-fuzzy method; Section 5.4.1.3) is the sum of the steps performed by the three tasks. In task 1, the time complexity for generating the reduct of the initial fuzzy rule is $O(nm)$, where n is the number of features. In task 2, the time complexity is $O(m^2)$. In task 3, the time complexity is $O(m)$.

Note that for the fourth phase (i.e., predicting adjustment through reasoning and adaptation; Section 5.4.1.4) the estimated solution for each case is calculated in the first step of the case selection process of phase 5 (i.e., step 1 of the case selection algorithm). This is shown below, where the computational effort is explained together with that required for the other steps in phase 5. In phase 5 (i.e., selecting representative cases; Section 5.4.1.5), the selection algorithm requires two major computations. The first step involves computing the case coverage in each cluster; on average this requires $(m/M)(m/M - 1)m^*p^*$ operations, where m is the number of cases, M the number of clusters, m^* the average number of fuzzy adaptation rules for each case, and p^* the average number of antecedents in each rule. Since m^* and p^* are very small compared with m, the complexity of step 1 in the selection algorithm is $O(m^2)$. Steps 2 and 3 involve sorting the cases according to their coverage ability, and the computation complexity will not exceed $O(m^2)$. Therefore, the overall computation complexity of the selection algorithm in this phase is also bounded by $O(m^2)$.

The space complexity is determined by the size of the case base and the temporary storage required in each phase described in Section 5.4.1. Among these phases, the multiplication of the two matrices in phase 2 requires the largest amount of memory (i.e., m^2). Since hardware and memory cost have been reduced significantly recently, memory space is not a critical concern when using this approach.

5.5   CASE-BASE MAINTENANCE USING A FUZZY INTEGRAL APPROACH

In Section 5.4, we have presented a rough-fuzzy approach for building and maintaining a client-side case base. In this section we describe another case-base maintenance (CBM) method, where the fuzzy integral technique is applied for modeling the competence of a given CBR system. Case-base competence has been brought sharply into focus since many maintenance policies are linked directly with the heuristics of measuring case-base competence to guide the maintenance procedures [33,34,51–53]. However, most of the current competence heuristics provide only coarse-grained estimates of competence. For example, Smyth and McKenna [54–59] and Smyth and McClave [60] employed a case deletion policy guided by a category-based competence model, where the cases are classified in only four basic competence categories. Zhu and Yang [61] provided a case addition policy based on the concept of case neighborhood, which is a coarse approximation of case coverage. Modeling case-base competence becomes a crucial issue in the field of CBM.

Smyth and McKenna proposed a competence model, where the concept of competence group is introduced due to the overlapping of individual case coverage and is defined in such a way that there are no overlaps within the case coverage in different competence groups. Group size and density have been considered in the definition of group coverage. Then the overall case-base competence can be computed simply by summing up the coverage of each group. However, the distribution of each group is not taken into account in this model. More specifically, it always assumes that the distribution of cases in each group is uniform, which sometimes leads to over- or underestimation of case-base competence. This problem is addressed by adopting a fuzzy integral–based competence model to compute the competence of a given CBR system more accurately. Consider a competence group; we first repartition it to ensure that the distribution of cases in each newly obtained group is nearly uniform. Since there are overlaps among the coverage of different new groups, fuzzy measures (nonadditive set functions) can be used to describe these overlaps because of their nonadditive characteristics. Fuzzy integrals are the corresponding integrals with respect to these fuzzy measures. As the most important tools of aggregation in information fusion, fuzzy integrals are appropriate here for computing the overall case-base competence of a CBR system. Some type of fuzzy measure, together with the corresponding fuzzy integral, the Choquet integral [62–64], are adopted in this approach. Next we describe briefly fuzzy measures and fuzzy integrals.

5.5.1 Fuzzy Measures and Fuzzy Integrals

The traditional tool of aggregation for information fusion is the *weighted-average method*, which is essentially a linear integral. It is based on the assumption that the information sources involved are noninteractive, and hence their weighted effects are viewed as additive. This assumption is not realistic in many applications. To describe the interaction among various information sources in such cases, a new mathematical tool, fuzzy measures or nonadditive set functions, can be used instead. In other words, a nonlinear integral such as the Choquet integral with respect to the nonadditive set functions can be used instead of the classical weighted average for information fusion.

More formally, fuzzy measures and fuzzy integrals are defined as follows: Let X be a nonempty set and $P(X)$ be the power set of X. We use the symbol μ' to denote a nonnegative set function defined on $P(X)$ with the properties $\mu'(\emptyset) = 0$. If

$\mu'(X) = 1, \mu'$ is said to be *regular*. It is a generalization of classic measure. When X is finite, μ' is usually called a *fuzzy* measure if it satisfies monotonicity; that is,

$$A \subseteq B \Rightarrow \mu'(A) \leq \mu'(B) \qquad \text{for A}, B \in P(X)$$

For a nonnegative set function μ', there are some associated concepts. For $A, B \in P(X), \mu'$ is said to be *additive* if $\mu'(A \cup B) = \mu'(A) + \mu'(B)$, to be *subadditive* if $\mu'(A \cup B) \leq \mu'(A) + \mu'(B)$, and to be *superadditive* if $\mu'(A \cup B) \geq \mu'(A) + \mu'(B)$. If we regard $\mu'(A)$ and $\mu'(B)$ as the importance of subsets A and B, respectively, the additivity of the set function means that there is no interaction between A and B; that is, the joint importance of A and B is just the sum of their respective importance. Superadditivity means that the joint importance of A and B is greater than or equal to the sum of their respective importance, which indicates that the two sets are enhancing each other. Subadditivity means that the joint importance of the two sets A and B is less than or equal to the sum of their respective importance, which indicates that the two sets are resisting each other.

Due to the nonadditivity of the fuzzy measures, some new types of integrals (known as *fuzzy integrals*), such as the Choquet integral, Sugeno integral, and N-integral, are used. Here we give only a definition of the Choquet integral, which is used in this section: Let $X = \{x_1, x_2, \ldots, x_n\}, \mu'$ be a fuzzy measure defined on the power set of X, and f be a function from X to [0,1]. The Choquet integral of f with respect to μ' is defined by

$$(C) \int f \, d\mu' = \sum [(f(x_i) - f(x_{i-1})]\mu'(A_i)$$

where we assume without loss of generality that $0 = f(x_0) \leq f(x_1) \leq \cdots \leq f(x_n)$ and

$$A_i = \{x_i, x_{i+1}, \ldots, x_n\}.$$

Next, we give an example [65] to illustrate how fuzzy measure and fuzzy integral describe the interactions of different objects. Let there be three workers, a, b, and c, working for $f(a) = 10, f(b) = 15$, and $f(c) = 7$ days, respectively, to manufacture a particular type of product. Without a manager, they begin work on the same day. Their efficiencies of working alone are 5, 6, and 8 products per day, respectively. Their joint efficiencies are not the simple sum of the corresponding efficiencies given above, but are as follows:

Workers	Products/Day
{a,b}	14
{a,c}	7
{a,c}	16
{a,b,c}	18

These efficiencies can be regarded as a fuzzy measure, μ', defined on the power set of $X = \{a,b,c\}$ with $\mu'(\emptyset) = 0$ (meaning that there is no product if no worker is there). Here inequality $\mu'(\{a,b\}) > \mu'(\{a\}) + \mu'(\{b\})$ means that a and b cooperate well, while inequality $\mu'(\{a,c\}) < \mu'(\{a\}) + \mu'(\{c\})$ means that a and c have a bad relationship and are not suitable to work together. Here μ' can be considered as an efficiency measure. In such a simple manner, during the first 7 days, all workers work together with efficiency $\mu'(\{a,b,c\})$, and the number of products is $f(c)\mu'(\{a,b,c\}) = 7 \times 18 = 126$; during the next $f(a) - f(c)$ days, workers a and b work together with efficiency $\mu'(\{a,b\})$, and the number of products is $[f(a) - f(c)]\mu'(\{a,b\}) = 3 \times 14 = 42$; during the last $f(b) - f(a)$ days, only b works with efficiency $\mu'(\{b\})$, and the number of products is $[f(b) - f(a)]$ $\mu'(\{b\}) = 5 \times 6 = 30$.

The function f defined on $X = \{a,b,c\}$ is called an *information function*. Thus, the value of the Choquet integral of f with respect to μ' is

$$(C) = \int f\, d\mu' = f(c)\mu'(\{a,b,c\}) + [f(a) - f(c)]\mu'(\{a,b\})$$
$$+ [f(b) - f(a)]\mu'(b) = 198$$

the total number of products manufactured by these workers during these days. Note that the meaning of fuzzy measures and fuzzy integrals is problem dependent. In this section, they are used to describe the coverage contributions of cases in a case base.

Before the fuzzy integral–based competence model is explained, the closely related (nonfuzzy) competence model of Smyth and McKenna [33] is described for convenience, together with its limitations.

5.5.2 Case-Base Competence

Smyth and McKenna [33], Smyth [34], and Smyth and Keane [51] explained the concept of case-base competence, and subsequently, various concepts such as *coverage* and *reachability* for measuring the problem-solving ability of case bases were developed. Some statistical properties of a case base (e.g., the size and density of cases) are used as input parameters for modeling case-base competence. A concept of competence group, which implies that different groups have no interaction (overlap) with each other, is also given in Smyth [34] as the fundamental computing unit of case-base competence.

The *competence* of a group of cases (G) (i.e., *group coverage of G*) depends on the number of cases in the group and its density. This is defined as

$$\text{GroupCoverage}(G) = 1 + |G|(1 - \text{GroupDensity}(G)) \qquad (5.19)$$

Here GroupDensity is defined as the average CaseDensity of the group:

$$\text{GroupDensity}(G) = \sum_{e \in G} \text{CaseDensity}(e,G)/|G| \qquad (5.20)$$

where

$$\text{CaseDensity}(e, G) = \sum_{e^* \in G - \{e\}} \text{SM}(e, e^*)/(|G| - 1) \qquad (5.21)$$

and $|G|$ is the size of the competence group G (i.e., the number of cases in group G). Various ways of computing the similarity between two cases e and e^* [i.e., $\text{SM}(e, e^*)$] are explained in Chapter 3, depending on the problems at hand.

For a given case base CB, with competence groups $\{G_1, G_2, \ldots, G_n\}$, the total coverage or the case-base competence is defined by

$$\text{Coverage(CB)} = \sum_{G_i \in G} \text{GroupCoverage}(G_i) \qquad (5.22)$$

From equation (5.22) it is seen that the definition of case-base competence took only the concepts of group size and group density into account. However, the distribution of cases in a case base also influences the case-base competence. For example, consider Figure 5.8, where the cases in part (*b*) are distributed uniformly,

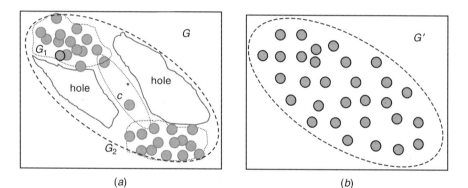

Figure 5.8 Examples of uniform and nonuniform distributions.

whereas those in parts (a) and (c) are not. Therefore, it is appropriate and necessary to incorporate this while computing the case-base competence of a case base.

Moreover, this model assumes that there is no overlap among different competence groups of cases (e.g., features interaction [66] is a common cause of overlaps). Therefore, simply by taking the sum of group competences as the overall case-base competence, without considering the overlapping effects, the resulting group competence may be over- or underexaggerated. This group overlap problem has been tackled by Shiu et al. [40] using fuzzy integrals. Details are given below.

5.5.3   Fuzzy Integral–Based Competence Model

Consider Figures 5.8a and c where we can easily see that cases such as c^* and c^{**} play an important role in affecting the overall competence distribution in the group. Therefore, it is important to detect such cases (which are called *weak links* in the following discussion) for possible identification of smaller competence groups, such as G_1, G_2, and G_3 in Figure 5.8c, those are having more evenly distributed cases. The competence of these smaller groups can then be computed using equations (5.19) to (5.21). It is worth noting that the competence of weak links can be considered to be their respective individual coverage, which reflects the relation among the several new groups. A new way of computing group competence based on this principle is described next.

5.5.3.1 Competence Error   In general, competence groups such as G_1 and G_2 in Figure 5.8a do not necessarily have the same strictly uniform distribution, and the weak-link case c^* is not necessarily a pivotal case (a case that cannot be solved by other cases). To deal with this situation, GroupDensity(G_1) [which is assumed to be equal to GroupDensity(G_2)] can be replaced by the average group density of groups G_1 and G_2, which can be denoted by $\overline{\text{GroupDensity}(G_i)}, i \in \{1,2\}$. Let $[\overline{\text{GroupDensity}(G_i)} - \text{GroupDensity}(G)]$ be denoted by ΔGroupDensity. The concept of *quasi-uniform distribution* can be used to describe the case-base distributions that are close to uniform distribution. As mentioned, the other assumption—that c^* is a pivotal case in the example—is not necessarily true in many cases. To address this problem, just consider the individual competence of c^* as its relative coverage, which is defined as

$$\text{RelativeCoverage}(e) = \sum_{e' \in \text{CoverageSet}(e)} \frac{1}{|\text{ReachabilitySet}(e')|} \qquad (5.23)$$

Then define

Competence Error(c^*)

$$= |G|\Delta\text{GroupDensity} - \overline{\text{GroupDensity}(G_i)} - \text{RelativeCoverage}(c^*)$$

$$\geq |G|\Delta\text{GroupDensity} - (\text{RelativeCoverage}(c^*) + 1) \qquad (5.24)$$

Since RelativeCoverage(c^*) is small, we can see that it is ΔGroupDensity that leads primarily to competence error.

5.5.3.2 Weak-Link Detection

To tackle the problem of nonuniformly distributed cases, it is necessary first, as noted earlier, to identify the weak links in each competence group. Definitions of weak link and several other concepts that are related more directly to the competence of the group in question, are provided below.

Definition 5.8. Let $G = \{G_1, G_2, \ldots, G_n\}$ be a set of competence groups in a case base CB. $c^* \in G$ is called a *weak link* if Competence Error $(c^*) \geq \alpha$, where α is a parameter that is defined by the user, depending on the requirement. If $\exists\, c^* \in G, c^*$ is a *weak link*, the competence group G is called a *nonuniform distributed competence group*. Otherwise, if $\forall\, e \in G$, Competence Error$(e) \leq \alpha$, then G is called *a quasi-uniform distributed competence group*.

A recursive method is used here to detect the weak links in a given competence group G, as follows:

Weak-Link Detection Algorithm

Step 1. W-SET $\leftarrow \{\cdot\}$, G-SET $\leftarrow \{\cdot\}$, $i = |G|$.

Step 2. if $(i \neq 0)$ {Consider each given competence group G in the competence model of Section 5.5.2, and compute Competence Error $(e), \forall\, e \in G; i = i - 1$}.

Step 3. If there is no weak link, add G to G-SET, end.

Step 4. If there is a weak link c^*, identify the competence groups G_1, $G_2, \ldots, G_n (n \geq 1)$ in $G - \{c^*\}$ using the competence model in Section 5.5.2; add c^* to the set of weak links W-SET.

Step 5. For $(1 \leq i \leq n)\{G \leftarrow G_i$; repeat steps 1 to 4}.

Thus, we can obtain the set of weak links W-SET in a given competence group G and the set of new competence groups G-SET.

5.5.3.3 Overall Coverage of Competence Group Using Fuzzy Integral

After detecting the weak links in a competence group G and cutting them off, let n new competence groups $G_1, G_2, \ldots, G_n (n \geq 1)$ be produced. According to the definition of a weak link, each newly produced group is sure to be quasi- uniformly distributed. The next task is to compute the overall coverage or competence of G. In the example described in Figure 5.8a, the overall competence of G can be calculated simply by the sum of the competence of $G_i (1 \leq i \leq n)$ and the relative coverage of c^*, but this method is not representative. There could be more complicated situations, as illustrated in Figure 5.8c, where it is difficult to identify clearly the contribution of each weak link. For example, in Figure 5.8c, c^* has much more influence on the coverage of G than c^{**} has, which reflects different relations among new competence groups. Therefore, a powerful tool, called a *fuzzy integral* (or nonlinear integral) with respect to a fuzzy measure (a nonadditive set function), is applied to describe this complex relationship.

Determining the λ-Fuzzy Measure μ' When the fuzzy integral is used to compute the overall coverage of the original competence group G, it is necessary first to determine the importance measure μ' of the n small competence groups $G_i(1 \leq i \leq n)$ both individually and in all possible combinations, the total number being $(2^n - 1)$. For cases in Figure 5.8c, there will be seven values of such measure:

$$\mu'(G_1), \mu'(G_2), \mu'(G_3), \mu'(G_1 \cup G_2), \mu'(G_1 \cup G_3), \mu'(G_2 \cup G_3), \mu'(G_1 \cup G_2 \cup G_3)$$

where μ' values of the unions of small competence groups can be computed by the λ-fuzzy measure [62], which takes the following form:

$$\mu'(A \cup B) = \mu'(A) + \mu'(B) + \lambda\mu'(A)\mu'(B), \quad \lambda \in (-1, \infty) \tag{5.25}$$

If $\lambda \leq 0, \mu'$ is a subadditive measure; if $\lambda \geq 0, \mu'$ is a superadditive measure; if and only if $\lambda = 0, \mu'$ is additive. So the focus of determining the λ-fuzzy measure μ' falls on the determination of the importance of each single group and λ. Note that $\lambda \geq 0$ in this example.

Given $\mu'(G_i) = 1(1 \leq i \leq n)$, the main problem in computing μ' here is therefore to determine the parameter λ. It is obvious that the properties of the weak links between two groups are important for determining λ. In this model, coverage of a group refers to the area of the target problem space covered by the group. In this sense, the value of λ is closely related to the coverage of weak links and the density of their coverage sets. Consider two arbitrary new groups G_i and G_j. Let the W-SET between them be $C^* = \{c_1^*, \ldots, c_h^*\}$. Coverage$(C^*)$ and Density(C^*) are defined as follows:

$$\text{Coverage}(C^*) = \sum_{i=1}^{h} \text{RelativeCoverage}(c_i^*)$$

$$\text{Density}(C^*) = \sum_{i=1}^{h} \text{GroupDensity}(Cov(c_i^*))/h$$

where $Cov(c_i^*)$ is the coverage set of the ith weak link c_i^* between G_i and G_j. The coverage contribution of $G_i \cup G_j$ must be directly proportional to Coverage(C^*) and inversely proportional to Density(C^*). With these assumptions, the parameter λ is defined as

$$\lambda = \text{Coverage}(C^*)(1 - \text{Density}(C^*)) \tag{5.26}$$

The λ-fuzzy measure μ' of $(G_i \cup G_j)$ can then be determined using equation (5.25).

Using the Choquet Integral to Compute Competence (Coverage) Due to the non-additivity property of the set function μ', a new type of integral (known as *nonlinear integrals*) are used to compute the overall coverage of the original competence group G based on the μ' measures of the n constituting competence groups and their unions. A common type of nonlinear integral with respect to nonnegative monotone set functions is the Choquet integral [63]. Its use in computing the competence of group G is described below.

Let the competence group $G = \{G_1, G_2, \ldots, G_n\}$ be finite, where $G_1, G_2, \ldots, G_n$ are the new small competence groups as defined earlier. Let $f_i = $ Group Coverage (G_i) and the importance measure μ' satisfy

$$\mu'(G_i) = 1 (1 \le i \le n)$$
$$\mu'(A \cup B) = \mu'(A) + \mu'(B) + \lambda\mu'(A)\mu'(B)(\lambda \ge 0)$$

where λ is determined by Equation (5.25). The process of calculating the value of the Choquet integral is as follows:

Step 1. Rearrange $\{f_1, f_2, \ldots, f_n\}$ into a nondecreasing order such that

$$f_1^* \le f_2^* \le \cdots \le f_n^*$$

where $(f_1^*, f_2^*, \ldots, f_n^*)$ is a permutation of $(f_1, f_2, \ldots, f_n)$.
Step 2. Compute

$$\text{Coverage}(G) = \int f \, d\mu' = \sum_{j=1}^{n} (f_j^* - f_{j-1}^*)\mu'(\{G_j, G_{j+1}, \ldots, G_n\})$$

where $f_0^* = 0$.

The value of the Choquet integral provides the coverage of the competence group G being considered. For a case base with several competence groups, the sum of the coverage of the individual groups gives the overall coverage of the case base.

5.5.4 Experiment Results

In this section some empirical results are provided to demonstrate the effectiveness of the fuzzy integral method (Section 5.5.3) in closely matching the actual competence of a case base. At the same time, the model in Section 5.5.2 is shown not to be a good predictor when the case base is not uniformly distributed. For this purpose, a small case base containing 120 cases, each of dimension 2, is considered. Each case is chosen randomly so that the case base satisfies nonuniform distribution. During

TABLE 5.8   Comparison of the Three Competence Models

Index	True	Model in Section 5.5.2	Fuzzy Integral Model in Section 5.5.3
Density	—	0.4	0.6
Competence	34.5	49.6	38.9
Error_number	0	15.1	4.4
Error_percent	0	43.8	12.8

the investigation, 50 randomly chosen cases in the case base are used as unknown target problems; the remaining 70 cases are used to form the experimental case bases.

The success criterion used is a similarity threshold: If the system does not retrieve any cases within this threshold, a failure is announced. True competence is regarded as the number of problems solved successfully. The experiment was repeated 100 times. The results shown in Table 5.8 are the average values computed over these iterations. In the table, "error_percent" represents the relative error of coverage of a model with respect to the true model, and is defined as error_percent = error_number/true_competence.

As expected, the error_percent of the fuzzy integral model is rather lower than that of the model in Section 5.5.2. When the number of cases increases, the former can reduce the competence error strikingly compared to the latter. In this experiment, the case base considered has a nonuniform distribution, but in the situation of uniform distributed case bases, the fuzzy integral competence model can still be used because if there is no weak link, the competence computed by the fuzzy integral model will be the same as that obtained using the model in Section 5.5.2.

5.6   SUMMARY

In this chapter we first presented the concepts and different techniques of case-base maintenance. Then the use of fuzzy set, rough set, and fuzzy integral for maintaining a distributed case-base reasoning system is demonstrated along with some experimental demonstrations using a source datum from the travel domain. Different features are explained through examples. The application of these soft computing techniques has proven to be very useful, particularly in the modeling and extraction of the domain knowledge in the case base. Future research includes the study of other soft computing techniques, such as neural networks, genetic algorithms, and various hybrid approaches for the maintenance of CBR systems.

REFERENCES

1. U. G. Gupta, *Validating and Verifying Knowledge-Based Systems,* IEEE Computer Society Press, Los Alamitos, CA, 1991.

2. U. G. Gupta, Validation and verification of knowledge-based systems: a survey, *Journal of Applied Intelligence*, vol. 3, pp. 343–363, 1993.

3. R. E. O'Keefe and D. E. O'Leary, Expert system verification and validation: a survey and tutorial, *Artificial Intelligence Review*, vol. 7, pp. 3–42, 1993.

4. F. Coenen and C. T. Bench, *Maintenance of Knowledge-Based Systems*, Academic Press, San Diego, CA, 1993.

5. N. K. Liu, Formal verification of some potential contradictions in knowledge base using a high level net approach, *Applied Intelligence,* vol. 6, pp. 325–343, 1996.

6. S. Murrell and R. Plant, On the validation and verification of production systems: a graph reduction approach, *International Journal of Human–Computer Studies*, vol. 44, pp. 127–144, 1996.

7. T. Menzies, Knowledge maintenance: the state of the art, *Knowledge Engineering Review*, vol. 14, no. 1, pp. 1–46, 1999.

8. D. W. Aha, A proposal for refining case libraries, in *Proceedings of the Fifth German Workshop on Case-Based Reasoning: Foundations, Systems, and Applications (GWCBR-97)*, Kaiserslautern, Germany, pp. 11–20, 1997.

9. D. W. Aha and L. A. Breslow, Learning to refine case libraries: initial results, *Technical Report AIC-97-003*, Navy Center for Applied Research in AI, Washington, DC, pp. 9–15, 1997.

10. D. W. Aha and L. A. Breslow, Refining conversational case libraries, in *Proceedings of the Second International Conference on Case-Based Reasoning (ICCBR-97)*, Providence, RI, Springer-Verlag, Berlin, pp. 267–278, 1997.

11. M. Angela and B. Smyth, Distributing case-base maintenance: the collaborative maintenance approach, *Computational Intelligence*, vol. 17, no. 2, pp. 315–330, 2001.

12. H. M. Avila, Case-base maintenance by integrating case-index revision and case-retention policies in a derivational replay framework, *Computational Intelligence*, vol. 17, no. 2, pp. 280–294, 2001.

13. S. Fox and B. Leake, Learning to refine indexing by introspective reasoning, in *Proceedings of the Fourteenth International Joint Conference on Artificial Intelligence (IJCAI-95)*, Montreal, Quebec, Canada, Morgan Kaufmann, San Francisco, pp. 1226–1232, 1995.

14. I. Iglezakis and T. R. Berghofer, A survey regarding the central role of the case base for maintenance in case-based reasoning, in *Proceedings of the ECAI Workshop on Flexible Strategies for Maintaining Knowledge Containers*, Humboldt University, Berlin, pp. 22–28, 2000.

15. I. Iglezakis and C. E. Anderson, Towards the use of case properties for maintaining case based reasoning systems, in *Proceedings of the Pacific Knowledge Acquisition Workshop (PKAW-00)*, Sydney, Australia, pp. 135–146, 2000.

16. I. Iglezakis, T. R. Berghofer, and C. Anderson, The application of case properties in maintaining case-based reasoning systems, in *Proceedings of the Ninth German Workshop on Case-Based Reasoning (GWCBR-01)*, Baden-Baden, Germany, pp. 209–219, 2001.

17. T. Reinartz, I. Iglezakis, and T. R. Berghofer, On quality measures for case base maintenance, in *Proceedings of the Fifth European Workshop on Case-Based Reasoning (EWCBR-00)*, Trento, Italy, Springer-Verlag, Berlin, pp. 247–259, 2000.

18. T. Reinartz and I. Iglezakis, Review and restore for case-base maintenance, *Computational Intelligence*, vol. 17, no. 2, pp. 214–234, 2001.

19. L. Portinale, P. Torasso, and P. Tavano, Dynamic case memory management, in *Proceedings of the Thirteenth European Conference on Artificial Intelligence (ECAI-98)*, Brighton, East Sussex, England, pp. 73–77, 1998.

20. Z. Zhang and Q. Yang, Towards lifetime maintenance of case-based indexes for continual case-based reasoning, in *Proceedings of the Eighth International Conference on Artificial Intelligence: Methodology, Systems, Applications*, Sozopol, Bulgaria, pp. 489–500, 1998.

21. D. B. Leake and D. C. Wilson, Categorizing case-base maintenance: dimensions and directions, in *Proceedings of the Fourth European Workshop of Case-Based Reasoning (EWCBR-98)*, Dublin, Ireland, Springer-Verlag, Berlin, pp. 196–207, 1998.

22. S. K. Pal and D. K. Dutta Majumder, *Fuzzy Mathematical Approaches to Pattern Recognition*, Wiley (Halstead), New York, 1986.

23. S. K. Pal and S. Mitra, *Neuro-Fuzzy Pattern Recognition*, Wiley, New York, 1999.

24. A. Preece, Building the right system right, in *Proceedings of the AAAI-98 Workshop on Verification and Validation of Knowledge-Based Systems,* Madison, WI, AAAI Press, Menlo Park, CA, pp. 38–45, 1998.

25. D. E. O'Leary, Verification and validation of case-based systems, *Expert Systems with Applications*, vol. 6, pp. 57–66, 1993.

26. K. Racine and Q. Yang, On the consistency management of large case bases: the case for validation, in *Proceedings of the AAAI-96 Workshop on Knowledge Base Validation*, Portland, OR, AAAI Press, Menlo Park, CA, pp. 84–90, 1996.

27. K. Racine and Q. Yang, Maintaining unstructured case, in *Proceedings of the Second International Conference on Case Based Reasoning (ICCBR-97)*, Providence, RI, Springer-Verlag, Berlin, pp. 553–564, 1997.

28. K. Racine and Q. Yang, Redundancy detection in semistructured case bases, *IEEE Transactions on Knowledge and Data Engineering*, vol. 13, no. 3, pp. 513–518, 2001.

29. Q. Yang, K. Racine, and Z. Zhang, Acquisition and maintenance of text-based plans, in *Proceedings of the 1998 AIPS (Artificial Intelligence Planning Systems Conference) Workshop on Knowledge Engineering and Acquisition for Planning: Bridging Theory and Practice*, PA, Pittsburgh, PA, 1998. Also available as *AAAI Technical Report WS-98–03*, pp. 113–122, 1998.

30. D. B. Leake and D. B. Wilson, When experience is wrong: examining CBR for changing tasks and environments, in *Proceedings of the Third International Conference on Case-Based Reasoning (ICCBR-99)*, Seeon Monastery, Munich, Spring-Verlag, Berlin, pp. 218–232, 1999.

31. D. B. Leake and R. Sooriamurthi, When two cases are better than one: exploiting multiple case-bases, In *Proceedings of the Fourth International Conference on Case-Based Reasoning (ICCBR-01)*, Vancouver, British Columbia, Canada, Springer-Verlag, Berlin, pp. 321–335, 2001.

32. D. B. Leake, Computer Science Department at Indiana University, *http://www.cs.indiana.edu/hyplan/leake.html*.

33. B. Smyth and E. McKenna, Modeling the competence of case-bases, in *Proceedings of the Fourth European Workshop of Case-Base Reasoning (EWCBR-98)*, Dublin, Ireland, Springer-Verlag, Berlin, pp. 23–25, 1998.

34. B. Smyth, Case-base maintenance, in *Proceedings of the Eleventh International Conference on Industrial and Engineering Applications of Artificial Intelligence and*

Expert Systems (IEA/AIE-98), Castellón, Spain, Springer-Verlag, Berlin, vol. 2, pp. 507–516, 1998.

35. E. McKenna and B. Smyth, An interactive visualization tool for case-based reasoners, *Applied Intelligence: Special Issue on Interactive Case-Based Reasoning*, pp. 95–114, 2001.

36. E. McKenna and B. Smyth, Competence-guided editing technique learning, in *Proceedings of the Fourteenth European Workshop on Case-Based Reasoning (EWCBR-00)*, Trento, Italy, Springer-Verlag, Berlin, pp. 186–197, 2000.

37. E. McKenna and B. Smyth, Competence-guided editing methods for lazy learning, in *Proceedings of the Fourteenth European Conference on Artificial Intelligence (EWCBR-00)*, Trento, Italy, Springer-Verlag, Berlin, pp. 553–564, 2000.

38. C. K. Simon, S. C. K. Shiu, C. H. Sun, X. Z. Wang, and D. S. Yeung, "Maintaining case-based reasoning systems using fuzzy decision trees, in *Proceedings of the Fifth European Workshop on Case-Based Reasoning (EWCBR-00)*, Trento, Italy, Springer-Verlag, Berlin, pp. 285–296, 2000.

39. S. C. K. Shiu, C. H. Sun., X. Z. Wang, and D. S. Yeung, Transferring case knowledge to adaptation knowledge: an approach for case-base maintenance, *Computational Intelligence*, vol. 17, pp. 295–314, 2001.

40. S. C. K. Shiu, Y. Li, and X. Z. Wang, Using fuzzy integral to model case-base competence, in *Proceedings of Soft Computing in Case-Based Reasoning Workshop in Conjunction with the Fourth International Conference in Case-Based Reasoning (ICCBR-01)*, Vancouver, British Columbia, Canada, Springer-Verlag, Berlin, pp. 206–212, 2001.

41. C. Hayes, P. Cunningham, and M. Doyle, Distributed CBR using XML, *http://wwwagr.informatik.uni-kl.de/~lsa/CBR/wwwcbrindex.html*.

42. L. McGinty and B. Smyth, Collaborative case-based reasoning: applications in personalized route planning, in *Proceedings of the Fourth International Conference on Case-Based Reasoning (ICCBR-01)*, Vancouver, British Columbia, Canada, Springer-Verlag, Berlin, pp. 362–376, 2001.

43. E. Plaza and S. Ontañón, Ensemble case-based reasoning: collaboration policies for multiagent cooperative CBR, in *Proceedings of the Fourth International Conference on Case-Based Reasoning (ICCBR-2001)*, Vancouver, British Columbia, Canada, Springer-Verlag, Berlin, pp. 437–451, 2001.

44. E. Plaza, J. L. Arcos, and F. Martin, Cooperative case-based reasoning, *Lecture Notes in Artificial Intelligence*, vol. 1221, pp. 180–201, 1997.

45. M. V. N. Prasad, V. Lesser, and S. Lander, Retrieval and reasoning in distributed case-bases, *Journal of Visual Communication and Image Representation, Special Issue on Digital Libraries*, vol. 7, no. 1, pp. 74–87, 1996.

46. S. C. K. Shiu, Department of Computing at the Hong Kong Polytechnic University, *http://www.comp.polyu.edu.hk/~csckshiu*.

47. T. Kohonen, *Self-Organization and Associate Memory*, Springer-Verlag, Berlin, 1988.

48. G. Fu, An algorithm for computing the transitive closure of a similarity matrix, *Fuzzy Sets and Systems*, vol. 51, pp. 189–194, 1992.

49. X. Z. Wang and J. R. Hong, Learning optimization in simplifying fuzzy rules, *Fuzzy Sets and Systems*, vol. 106, pp. 349–356, 1999.

50. Y. Yuan and M. J. Shaw, Induction of fuzzy decision trees, *Fuzzy Sets and Systems*, vol. 69, pp. 125–139, 1995.

51. B. Smyth and M. T. Keane, Remembering to forget: a competence-preserving case deletion policy for case-based reasoning systems, in *Proceedings of the Fourteenth International Joint Conference on Artificial Intelligence (IJCAI-95)*, Montreal, Quebec, Canada, Morgan Kaufmann, San Francisco, pp. 377–382, 1995.

52. Q. Yang and J. Zhu, A case-addition policy for case-base maintenance, *Computational Intelligence*, vol. 17, no. 2, pp. 250–262, 2001.

53. D. Leake and D. Wilson, Remembering why to remember: performance-guided case-base maintenance, in *Proceedings of the Fifth European Workshop on Case-Based Reasoning (EWCBR-00)*, Trento, Italy, Springer-Verlag, Berlin, pp. 161–172, 2000.

54. B. Smyth and E. McKenna, Building compact competent case-bases, in *Proceedings of the Third International Conference on Case-Based Reasoning (ICCBR-99)*, Seeon Monastery, Munich, Springer-Verlag, Berlin, pp. 329–342, 1999.

55. B. Smyth and E. McKenna, Footprint-based retrieval, in *Proceedings of the Third International Conference on Case-Based Reasoning (ICCBR-99)*, Seeon Monastery, Munich, Springer-Verlag, Berlin, pp. 343–357, 1999.

56. B. Smyth and E. McKenna, An efficient and effective procedure for updating a competence model for case-based reasoners, in *Proceedings of the Eleventh European Conference on Machine Learning*, Barcelona, Spring-Verlag, Berlin, pp. 357–368, 2000.

57. B. Smyth and E. McKenna, Incremental footprint-based retrieval, in *Proceedings of ES-00*, Cambridge, Spring-Verlag, Berlin, pp. 89–101, 2000.

58. B. Smyth and E. McKenna, Competence guided instance selection for case-based reasoning, in *Instance Selection and Construction: A Data Mining Perspective* (H. Liu and H. Motoda, (eds.), Kluwer Academic, Norwell, MA, pp.1–18, 2000.

59. B. Smyth and E. McKenna, Competence models and the maintenance problem, *Computational Intelligence*, vol. 17, no. 2, pp. 235–249, 2001.

60. B. Smyth and P. McClave, Similarity vs. diversity, in *Proceedings of the Fourth International Conference on Case-Based Reasoning (ICCBR-01)*, Vancouver, British Columbia, Canada, Springer-Verlag, Berlin, pp. 347–361, 2001.

61. J. Zhu and Q. Yang, Remembering to add: competence-preserving case addition policies for case base maintenance, in *Proceedings of the International Joint Conference in Artificial Intelligence (IJCAI-99)*, Stockholm, Sweden, Morgan Kaufmann, San Francisco, pp. 234–239, 1999.

62. Z. Y. Wang and G. J. Klir, *Fuzzy Measure Theory*, Plenum Press, New York, 1992.

63. E. Pap, *Null-Additive Set Function*, Kluwer Academic, Dordrecht, The Netherlands, 1996.

64. Z. Y. Wang, K. S. Leung, M. L. Wong, J. Fang, and K. Xu, Nonlinear nonnegative multiregressions based on Choquet integrals, *International Journal of Approximate Reasoning*, vol. 25, no. 2, pp. 71–87, 2000.

65. Z. Y. Wang, K. S. Leung, and J. Wang, A genetic algorithm for determining nonadditive set functions in information fusion, *Fuzzy Sets and Systems*, vol. 102, no. 3, pp. 463–469, 1999.

66. X. Z. Wang and D. S. Yeung, Using fuzzy integral to modeling case-based reasoning with feature interaction, in *Proceedings of the 2000 IEEE International Conference on Systems, Man, and Cybernetics*, Nashville, TN, vol. 5, pp. 3660–3665, 2000.

CHAPTER 6

APPLICATIONS

6.1 INTRODUCTION

As discussed in the previous chapters, soft computing techniques allow reasoning with uncertainty and imprecision, efficient learning of intractable classes (concepts), and robust parameter estimation through parallel searching. They are very useful for enhancing the overall problem-solving ability of CBR systems. In this chapter we describe briefly some successful soft CBR applications that are used in various domains, such as law, medicine, e-commerce, finance, oceanographic forecasting, and engineering. They provide advice and decision support to users by recalling previously successful cases for use as a guide or template. This chapter is organized as follows. In Section 6.2 we describe a fuzzy CBR application for Web access path prediction, where the concept of fuzzy set is used in case representation and mining association rules. In Section 6.3, a case-based medical diagnosis system incorporated with fuzzy neural network for case selection and adaptation is explained. Section 6.4 illustrates the application of a connectionist neural network for oceanographic forecasting. The use of fuzzy logic to capture the ambiguity of legal inference in a legal CBR system is explained in Section 6.5. This is followed, in Section 6.6, by a description of a residential property evaluation system that integrates CBR techniques with fuzzy preferences in determining similarities among cases. In Section 6.7 we describe an application using GAs to support bond rating and bankruptcy prediction. In Section 6.8, a color-matching system is described in which fuzzy logic is used in case

Foundations of Soft Case-Based Reasoning. By Sankar K. Pal and Simon C. K. Shiu
ISBN 0-471-08635-5 Copyright © 2004 John Wiley & Sons, Inc.

representation and matching. A CBR system in fashion footwear design is explained in Section 6.9, where a neuro-fuzzy technique is used in case indexing and case retrieval. In Section 6.10, other applications using soft computing techniques are mentioned.

6.2   WEB MINING

A fuzzy CBR application for predication of Web access patterns has been developed and tested by Wong et al. [1,2]. Here a fuzzy association rule mining algorithm, together with fuzzy case representation, is used to facilitate candidate case selection. With these rules, Web designers are able to interpret user interests and behavior. The personalization and prefetching of critical Web pages become possible. The method has experimentally demonstrated better prediction accuracy than some existing methods.

6.2.1   Case Representation Using Fuzzy Sets

Let a Web page be represented by a t-dimensional vector, where t is the number of permissible terms (i.e., features) in the Web page. Absence of a term is indicated by a zero, while the presence of a term is indicated by a positive number known as the *weight*. The *normalized weighting function* of a term in terms of its frequency of occurrence is defined as

$$w_{ij} = \frac{\mathrm{tf}_{ij} \cdot \mathrm{idf}_j}{\sqrt{\sum_{j=1}^{t} (\mathrm{tf}_{ij})^2 (\mathrm{idf}_j)^2}} \tag{6.1}$$

where w_{ij} is the weight of term j in Web page i, and tf_{ij} is the term frequency of term j in Web page i. idf_j is the inverse Web page frequency of term j in Web page i and is computed as

$$\mathrm{idf}_j = \log \frac{N}{n_j} \tag{6.2}$$

where N is the number of Web pages in the case base and n_j is the number of Web pages in the case base that contain the term t_j.

After computing all the term weights, they are used as the universe of discourse to formulate the linguistic variable *term weight importance*. Five fuzzy sets are defined for the term weight importance: highly frequent (HF), frequent (F), medium (M), less frequent (LF), and rare (R). A center-based membership function approach is used to determine a three-point triangular fuzzy membership function. The fuzzy c-means clustering algorithm is used to determine the centers of the triangular functions that use the term weights as the input data. The resulted shapes of the five membership functions are shown in Figure 6.1.

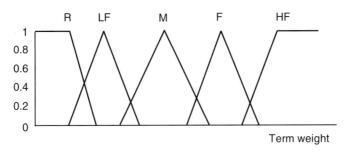

Figure 6.1 Fuzzy membership functions for the term weight.

6.2.2 Mining Fuzzy Association Rules

The relation between successive access paths of a user is regarded as sequentially associated. For example, if u_i is a user transaction with access sequence $\{a_1, a_2, \ldots, a_n\}$, Web access pages $a_i (i = 1, 2, \ldots, n)$ are then considered to be sequentially associated. The similarity between every two pages in a fixed order is computed based on a similarity measure defined on the features extracted. The similarity values are stored in cases as shown in Table 6.1.

6.2.2.1 Data Fuzzification To discover the fuzzy association rules from the Web case base, similarity values of the access pages in a user transaction are fuzzified into five linguistic terms: highly similar (HS), quite similar (QS), medium (M), not so similar (NSS), and not similar (NS). The process of determining the five membership functions is similar to that used in case representation (Section 6.2.1).

6.2.2.2 Mining Fuzzy Association Rules A fuzzy association rule is defined follows:

$$\text{IF } X = \{x_1, x_2, \ldots, x_n\} \text{ is } A = \{f_1, f_2, \ldots, f_{n-1}\}$$
$$\text{THEN } Y = \{y_1, y_2, \ldots, y_n\} \text{ is } B = \{g_1, g_2, \ldots, g_{n-1}\},$$

where X is the sequence of URLs accessed and A is the associated fuzzy set and Y is the sequence of URLs predicted and B is the associated fuzzy set. X is the problem part of the transaction u_i in the case base and Y is the corresponding solution part.

TABLE 6.1 Similarity Values of Web Pages

Case	$SM(a_1, a_2)$	$SM(a_2, a_3)$	$SM(a_3, a_4)$	$\cdots$	$SM(a_i, a_{i+1})$	$\cdots$
1	87	89	90	$\cdots$	$\cdots$	$\cdots$
2	56	78	70	$\cdots$	$\cdots$	$\cdots$
3	75	83	80	$\cdots$	$\cdots$	$\cdots$

Therefore, the fuzzy association rule in the form "IF the interpage similarity of X is A THEN the interpage similarity of Y is B" is to be mined. A and B represent the fuzzy sets associated with the corresponding accesses. Each fuzzy set, called an *item*, represents the linguistic term describing the similarity between two Web pages. As in a binary association rule, "X is A" is called the *antecedent* of the rule, and "Y is B" is called the *consequent* of the rule. In the traditional sense, for a rule to be interesting, it needs to have enough support and a high confidence value. To minimize the set of resulting fuzzy association rules, only the items that have transaction support above a user-specified threshold are used. Item sets with minimum support are called *frequent item sets*.

Let us now examine methods of computing fuzzy support and fuzzy confidence values. The fuzzy support value is computed by first summing the transactions of all the items in the specified item sets, then dividing it by the total number of records. A fuzzy support value reflects not only the number of records supporting the item set, but also their degree of support. The fuzzy support value of the item sets of $\langle X, A \rangle$ is computed as

$$FS_{\langle X,A \rangle} = \frac{\sum_{u_i \in CB} \prod_{x_j \in X, f_j \in A} u_i : (x_j, f_j)}{|CB|} \qquad (6.3)$$

where $|CB|$ is the cardinality of the case set.

The following example is used to illustrate the computation of the fuzzy support value. Let $X = \{URL_1, URL_2\}$ and $A = \{\text{highly Similar, medium}\}$. Some typical cases are shown in Table 6.2. Using the cases in the table, the fuzzy support value of the rule "IF the interpage similarity of URL_1 and URL_2 is highly similar THEN the interpage similarity of URL_2 and URL_3 is medium" is calculated [using equation (6.3)] as

$$FS\langle X,A \rangle = \frac{(0.5)(0.8) + (0.6)(0.6) + (0.4)(0.8) + (0.7)(0.2)}{4} = 0.305$$

The frequent item sets are used to generate all possible rules. If the union of an antecedent and a consequent has enough support, and the rule has high confidence, this rule is considered interesting. When a frequent item set $\langle Z, D \rangle$ is obtained, fuzzy association rules of the form "IF X is A THEN Y is B" are generated, where

TABLE 6.2 Cases with Membership Values

$\langle$Interpage Similarity of URL_1 and URL_2, Highly Similar$\rangle$	$\langle$Interpage Similarity of URL_2 and URL_3, Medium$\rangle$
0.5	0.8
0.6	0.6
0.4	0.8
0.7	0.2

$X \subset Z$, $Y = Z - X$, $A \subset D$, $B = D - A$, $\langle X, A \rangle$ is the antecedent, and $\langle Y, B \rangle$ is the consequent.

The fuzzy confidence value is obtained as follows:

$$
\begin{aligned}
\mathrm{FC}_{\langle\langle X,A\rangle,\langle Y,B\rangle\rangle} &= \frac{\mathrm{FS}_{\langle Z,D\rangle}}{\mathrm{FS}_{\langle X,A\rangle}} \\
&= \frac{\sum_{u_i \in C_s} \prod_{z_j \in Z, d_j \in D} u_i : (z_j, d_j)}{\sum_{u_i \in C_s} \prod_{x_j \in X, f_j \in A} u_i : (x_j, f_j)}
\end{aligned}
\tag{6.4}
$$

where $Z = X \cup Y$ and $D = A \cup B$. Since the fuzzy confidence value is a measure of the degree of support given by the transactions, it is also used to estimate the interestingness of the generated fuzzy association rules. In the equation, the fuzzy support of $\langle Z,D \rangle$ is divided by the fuzzy support of $\langle X,A \rangle$. Using the cases in Table 6.2, the fuzzy confidence of the rule "IF the interpage similarity of URL_1 and URL_2 is highly similar THEN the interpage similarity of URL_2 and URL_3 is medium" is

$$
\mathrm{FC}\langle\langle X,A\rangle, \langle Y,B\rangle\rangle = \frac{0.4 + 0.36 + 0.32 + 0.14}{0.5 + 0.6 + 0.4 + 0.7} = 0.555
$$

As a result, the fuzzy sequential association rules are mined whose fuzzy support and fuzzy confidence are above a given threshold. A set of candidate cases that satisfy the fuzzy rules is selected from the case base. Based on the assumption that people accessing similar page content will have similar access paths, these candidate cases are then used to predict the user Web access patterns and recommend the Web pages to be prefetched.

6.3 MEDICAL DIAGNOSIS

To help less-experienced physicians, Hsu and Ho [3,4] developed a hybrid CBR system incorporating fuzzy logic, neural networks, and induction technique to facilitate medical diagnosis. Details are given below.

6.3.1 System Architecture

The system consists of a case base, a commonsense knowledge base, a general adaptation knowledge base, a user interface, a data analyzer, a specification converter, and a case-based reasoner (see Fig. 6.2). In this system, each instance of the medical diagnosis is called a case. It has three parts: (1) a diagnosis description describing the scenario of how a diagnosis is processed; (2) a patient description describing both the subjective and objective findings, the pathology and the laboratory testing results of a patient; and (3) some domain-specific knowledge (i.e., case-specific adaptation rules) for case adaptation.

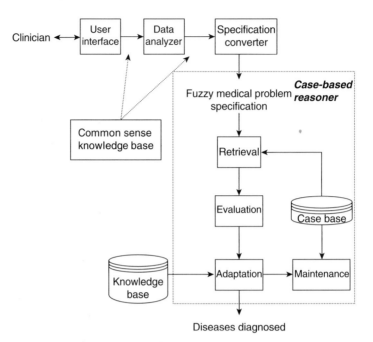

Figure 6.2   Structure of hybrid medical diagnosis system.

The user interface allows physicians to interact with the system by means of a friendly human body image. In the data analyzer, a medicine-related commonsense knowledge base is used to filter unreasonable input as well as to select the relevant features. The specification converter fuzzifies the significant features: the important patient data items along with the corresponding fuzzy degrees. In the case retrieval process, a distributed neural network is used to perform approximate matching. In case evaluation and adaptation, the constrained induction tree is used to determine the valuable features for adaptation. The neuro-fuzzy method used in this system is explained in the next section. This is followed by the method of case evaluation and adaptation using induction tree.

6.3.2   Case Retrieval Using a Fuzzy Neural Network

A distributed fuzzy neural network is used to select the most relevant cases to provide candidate solutions for a given input. It contains two layers (see Fig. 6.3). The first layer, the *symptom net*, determines the fuzzy similarity of the subjective and objective symptoms between the query patient case and the cases in the case base. Similar to the symptom net, the pathology net is used to compute the fuzzy similarity of the pathology and laboratory data between the query patient case and the cases in the case base. Each layer is further divided into subnets according to the pathological types of the cases: congenital type, neoplasm type, infection type, obstructive type, and noncongenital type [5]. The principal advantage of the subnet

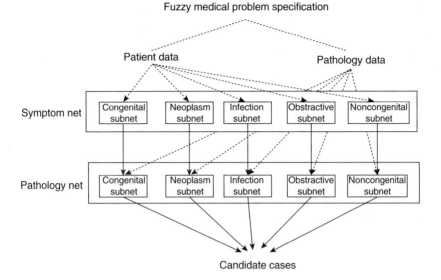

Figure 6.3 Distributed fuzzy neural networks for case retrieval. (From [3].)

design is to reduce the training load. When a new case is added to the case base, the system only needs to retrain the corresponding subnet(s) while keeping the others unchanged. Using this fuzzy neural network, one or more candidate cases are selected to provide potential solutions.

6.3.3 Case Evaluation and Adaptation Using Induction

In the evaluation phase, an induction tree is used to identify the relevant features in the candidate case(s). The nodes in the induction tree include both feature nodes and case nodes. Each feature node is branched into several child nodes, corresponding to different feature values. A case node is reached at the end of each branch. For example, for the candidate cases listed in Table 6.3, the induction tree shown in Figure 6.4 is developed. The significance of each feature is described by the expected utility (EU) for each node in the induction tree. Set $\text{EU}(e_i) = U(e_i)$, where $U(e_i)$ is the utility of case i computed by taking into account its adaptability [6]. Then the EU values of the father nodes of e_i (i.e., all the feature nodes) are computed recursively as

$$\text{EU}(F_j) = \sum_{k=1}^{n} \text{EU}(F_{jk})P(F_j \rightarrow F_{jk}) \tag{6.5}$$

where F_{jk} is the kth child node of feature node F_j, n the number of children of feature node F_j, and $P(F_j \rightarrow F_{jk})$ the probability of having the feature value F_{jk} given the feature F_j. $P(F_j \rightarrow F_{jk})$ is determined by the domain experts. It represents the probability of occurrence of the feature value F_{jk} in the case base. The more

TABLE 6.3 Example Candidate Cases

Features	Case 1 (e_1)	Case 2 (e_2)
Disease category (DC)	Respiratory	Respiratory
Disease organ (DO)	Pulmonary	Pulmonary
Practical illness 1 (PI1)	Cough	Cough
Practical illness 2 (PI2)	Dyspnea	Dyspnea
Effusion protein (EP)	High	High
Specific gravity (SG)	High	High
Chief complaint 1 (CC1)	Chest pain	Chest pain
Diagnostic 1 (DI1)	Pleural effusion	Pleural effusion
Diagnostic 2 (DI2)	Pneumococcal (DI2-1)	Streptococci pneumonia (DI2-2)
Cytology (CY)	Liver tumor (CY1)	Normal (CY2)
Diagnostic procedure 1 (DP1)	Pleural biopsy	Pleural biopsy
Diagnostic procedure 2 (DP2)	Pleurosocopy	Pleurosocopy

Source: [3].

frequently a feature value occurs in the case base, the higher the probability that it is relevant to the new problem.

Subsequently, with the help of adaptation knowledge (including case-specific adaptation rules and the general knowledge base), the corresponding adaptation strategy can be determined. For those features whose relevance is above a given threshold, no adaptation is needed. On the other hand, for those features whose relevance is below a given threshold, adaptation is carried out to create the final diagnosis for the patient.

In general, artificial neural networks (ANNs) are useful in medical case matching, selection, and retrieval, especially when there is insufficient domain knowledge or relevant information. However, the user needs to consider the resources required for successful training of the ANNs. Other approaches, such as k-NN (k-nearest neighbor) and RBS (rule-based system), may be feasible alternatives if the domain knowledge is or can be made available.

Another example of soft CBR application in medicine is the ELSI (error log similarity index) system [7,8]. It was developed to find the error patterns automatically using diagnostic information from computed tomography (CT) scanners. The results are categorized and stored as cases that could be used by

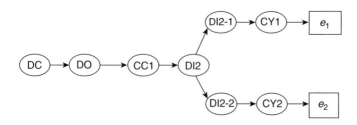

Figure 6.4 Example of the induction tree.

less experienced service engineers for control of CT scanners. In this application, the error log information is not well formatted and often contains erroneous messages. A membership function is introduced to describe the confidence of the case matching. The use of fuzzy logic has been shown to increase accuracy and reduce maintenance cost.

6.4 WEATHER PREDICTION

In 2001, Corchado and Lees [9] presented a *universal forecasting model*. It is a hybrid system in which a neural network is integrated within the CBR frame to provide improved case adaptation performance. Here *universal* means the ability to produce accurate results anywhere on any ocean at any time. The structure of the model and the case adaptation method using ANN are described below in brief.

6.4.1 Structure of the Hybrid CBR System

The hybrid system is composed of a CBR system and a radial basis function ANN (RBFANN). It is capable of adapting itself in real time to different oceanographic water masses. The experiments were carried out using data sets recorded in the Atlantic Ocean (cruises AMT 4), which include the sea-surface temperature recorded in real time by sensors in the vessels and satellite pictures received weekly. Figure 6.5 shows the detailed information flow throughout the CBR cycle, and especially how the ANN has been integrated with CBR operation to form a hybrid forecasting system.

6.4.2 Case Adaptation Using ANN

In the case adaptation algorithm, the training time required for the RBFANN is very short and no human intervention is required. This network obtains the most representative solution from a number of retrieved cases. Only a small number of rules supervise the training of the ANN, and it can learn without forgetting by adapting its internal structure (adding or deleting its centers). A case in the case base is described by a set of temperature values (called a *feature vector*). According to the similarity measure defined on the feature vectors, k best matches to the problem case can be retrieved using the k-NN algorithm (Section 3.2.5). These k cases will be used to train the RBFANN in the adaptation stage. Every time that the ANN is retrained, its internal structure is adapted to the new problem and the cases are adapted to produce the solution, which is a generalization of those cases. The parameter values of the RBFANN are shown in Table 6.4. This ANN uses nine input neurons, 20 to 35 neurons in the hidden layer, and one neuron in the output layer. Initially, 20 feature vectors chosen randomly from the training data set (i.e., the set of retrieved cases) are used as the centers of the radial basis functions used in the hidden layer of the ANN. The number of feature vectors depends on the data sets extracted in the AMT cruises [10] and changes during training. The topology of the

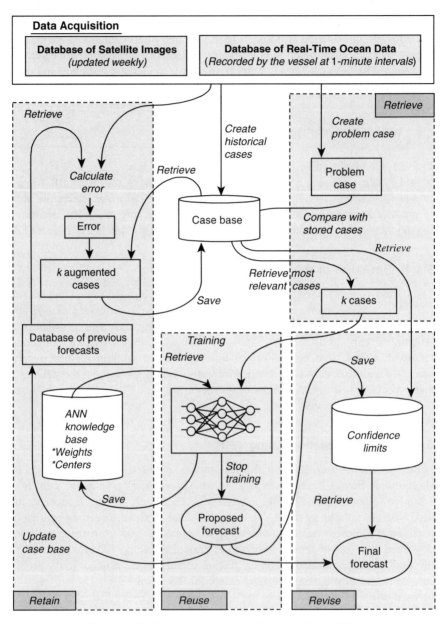

Figure 6.5 Structure of the hybrid system. (From [9].)

ANN (i.e., the number of neurons in each layer) is determined empirically before-hand.

To make the centers as close as possible to many vectors from the input space, the center and weight must be adapted. This type of adaptation is particularly important because of the high dimensionality of the input layer.

TABLE 6.4   Parameter Values of the Radial Functional Basis ANN

Number of input neurons	9
Number of neurons in the hidden layer	20–35
Number of neurons in the output layer	1
Input data	The difference between the present temperature values and those of the input profile taken every 4 km
Output data	The difference between the temperature at the present point and the temperature 5 km ahead

6.4.2.1 *Center Adaptation*

All the centers are associated with a Gaussian function, the width of which for all the functions is set as the Euclidean distance between the two centers that are separated the most from each other. The closest center to a particular input vector is moved toward the input vector by a percentage α of the present distance between them. α is initialized to 20 every time that the ANN is retrained, and its value is decreased linearly with the number of iterations until α becomes 0; then the ANN is trained for a number of iterations (e.g., between 10 and 30 iterations for the entire training data set, depending on the time left for the training) in order to obtain the most satisfied weights for the final value of the centers. The thresholds that determine the centers and weight adaptation are determined empirically.

6.4.2.2 *Weight Adaptation*

The delta rule [equation (B.9) in Appendix B] is used to adapt the weighted connections from the centers to the output neurons [11]. In particular, for each pair of input and desired output vectors presented, one adaptation step, according to the delta rule, is made. A new center can be inserted between the most distant center C (which can be determined by Euclidean distance) and the center closest to it when the average error in the training data set does not fall by more than 10% after 10 iterations (of the entire training set). Centers are also eliminated if the absolute value of the weight associated with a neuron is less than 20% of the average absolute value computed on the five smallest weights. The number of neurons in the middle layer is kept above 20. This is a simple and efficient way of reducing the size of the ANN without decreasing its memory dramatically.

After the adaptation of the ANN is complete, a more satisfactory solution (i.e., a crisp value) for the problem can be obtained, but in practice, this solution is hardly completely accurate. The next step is therefore to make some adaptation of the solution. Since this is a real-time problem, it is impossible to evaluate the outcome of the system before it is used. To improve this situation to some extent, an error limit can be defined to substitute the crisp output with a band (or error interval) around the output of the ANN. For example, if θ is the crisp solution obtained by the ANN and EL is the corresponding error limit, the error interval, defined as $[\theta - EL, \theta + EL]$, is used as the prediction instead of using the crisp value θ. If the error limits are too wide, the forecast will be meaningless. Therefore, a trade-off is made between a broad error limit (that will guarantee that the real solution is always within its bands) and the crisp solution.

For each water mass, a default error limit, EL_0, has been obtained empirically. Every time a cruise crosses a water mass, a new error limit EL_z (where $0 < z < 6$) is calculated by averaging the errors in all the predictions made. Therefore, there are at most five error limits associated with a water mass, which are used to determine the error limit for the present prediction. Note that this number of error limits is not critical. It is found that a smaller number can also guarantee stability, whereas a larger number may not provide a better result. The *average error* of each case retrieved, which is a measure of the average error for the previous predictions using this case, is also used in determining the error limit. The error limit determines the error interval centered in the crisp temperature value obtained by the ANN, and it is expected that there is a high probability that the forecast is within this interval. In this application, although the value of the probability varies depending on the distance of the forecast, it is required to be higher than 0.9.

The error limit, EL, is determined as shown below. If the output of the ANN is θ, the average value of the accumulated errors of the cases taking part in a given forecast is AE, and ACL is the average value of EL_z ($0 < z < 6$), then EL is computed by

$$EL = AE \times 0.65 + ACL \times 0.35 \tag{6.6}$$

The corresponding error interval is

$$[\theta - ((AE \times 0.65) + (ACL \times 0.35)), \theta + ((AE \times 0.65) + (ACL \times 0.35))]$$

The constant terms used in equation (6.6) were obtained empirically using a sufficient amount of data from all the water masses of the Atlantic Ocean.

Another methodology, combining CBR with fuzzy set theory to predict the weather, was developed by Bjarne and Denis [12] by determining the features of the cases that are significant in computing the similarity between cases. The knowledge is encoded in a similarity measure function and then used to retrieve k nearest neighbors (k-NN) from a large database. Prediction(s) for the query case are made from a weighted median of the outcomes of retrieved (past) cases. Past cases are weighted according to their degrees of similarity to the present case, which is described by fuzzy sets. Such a fuzzy k-NN-based prediction system, called WIND-1, is tested with the problem of producing six-hourly predictions of cloud ceiling and visibility in an airport, given a database of over 300,000 consecutive, hourly, airport weather observations (36 years of records). Its prediction accuracy is measured using standard meteorological statistics and compared with that of a benchmark prediction technique, persistence climatology (PC). In realistic simulations, WIND-1 is found to be significantly more accurate and efficient. It takes about only 1 minute to produce a forecast.

Several other approaches based on the CBR in conjunction with statistical techniques [13], with supervised ANN [14] and with unsupervised ANN [13], have been studied in the area of oceanographic foresting. The result of these methods suggests that to obtain accurate forecast in an environment in which the parameters are changing continuously, both temporally and spatially, a methodology is needed that can incorporate the strengths and abilities of several artificial intelligence approaches.

6.5 LEGAL INFERENCE

In this section a CBR legal system [15–17] is explained. Fuzzy logic is used to describe the fuzziness of the legal inference as well as for case representation and case matching. This system is used to teach the principles of contract laws in universities. The cases come from the United Nation's Convention on Contracts for the International Sale of Goods (CISG), which has been used in many countries. The system has four main components: case base, case retrieval engine, inference engine, and user interface.

6.5.1 Fuzzy Logic in Case Representation

In the legal case base, each case (or *precedent*) consists of a number of issues. Each issue has a set of features and case rules. It is stored as a frame, as follows:

```
Case n:
   ((Issue n₁)
        ⋮
     (Feature n₁)
        ⋮
     (Case rule n₁)
        ⋮
    (Issue n₂)
        ⋮
    (Issue nₘ)
        ⋮       )
```

Each issue consists of a legal argument point and a court judgment. It can be further interpreted (or categorized) into a number of features and case rules by law experts according to the statute rule and facts of the precedents. The case rules act as some specific knowledge to facilitate the legal inference. It can be interpreted as the connections between the precedents and the court judgments.

6.5.1.1 *Membership and Vagueness Values* To represent cases with uncertainty, membership values and vagueness values are being stored in frames called *fuzzy frames*. Each feature value in a case is described by a linguistic term

TABLE 6.5 Membership and Vagueness Values of the Linguistic Terms

Membership (m)	
Completely no (CN)	0
Probably no (PN)	0.25
More or less (ML)	0.5
Probably yes (PY)	0.75
Completely yes (CY)	1
Vagueness (v)	
Vague (V)	1
Roughly (R)	0.5
Clearly (C)	1

such as "completely no," "probably no," ..., or "completely yes." The linguistic term is represented by a membership value. The degree of uncertainty of this term is described by a vagueness value. The values of membership and vagueness can be assigned by the domain experts. For example, the corresponding relationships between the numerical representation and linguistic representation are shown in Table 6.5. There are five values for the membership and three values for the vagueness. Determining the number of values is problem dependent.

6.5.1.2 Example of Fuzzy Frame Representation
Assume that there is a situation description concerning an issue "The proposal is sufficiently definite" as follows:

Event: Proposal
Description of event:
    The goods are jet engine systems.
    The quantity of engine systems can be calculated by the quantity of planes that will be purchased.
Concerning the price:
    The price of Boeing jet engine is fixed.
    The jet engine system includes a support package, services, and so on.

With the help of available statutes and rules, the situation description for the given issue is first represented by a case in frame form. One of the statute rules, article 14 of the CISG, can be used to determine the relevant features for the issue and the judgment about whether the proposal is sufficiently definite. The statements in article 14 are: A proposal for concluding constitutes an offer if it is sufficiently definite and indicates the intention of the offerer to be bound in case of acceptance.

TABLE 6.6   Fuzzy Frame Case Representation

	N	
The proposal is sufficiently definite	*m*	*V*
The goods are indicated.	1.0	0.0
The quantity is fixed.	1.0	0.0
The entity price is fixed.	0.0	0.0
A part price is fixed.	1.0	0.0

A proposal is sufficiently definite if it indicates the goods and expressly or implicitly fixes or makes provision for determining the quantity and price.

According to article 14, the information that is relevant to goods, quantity, and price is considered to be important to represent the previous situation. Therefore, it can be characterized as follows:

Argument point: The proposal is sufficiently definite.

Judgment: No (N)

Features: The goods are indicated.

The quantity is fixed.

The entity price is not fixed.

A part price is fixed.

Further, it is represented by such a fuzzy frame, where the name of the frame is the argument point about a case, the frame value is the judgment about a case, the slot name is the feature about a case, and slot value has the values m and v. These are shown in Table 6.6. Similarly, the precedent in the case base can be represented by fuzzy frames. The case rules extracted from the relevant statutes and rules are also stored in a case. They are in the form of some questions providing to users, and they give a judgment about whether the candidate case(s) can be reused according to the question answers from the users. The case rule judgments are represented by fuzzy sets. The triangular membership functions are used.

6.5.2   Fuzzy Similarity in Case Retrieval and Inference

In the fuzzy frame representation of a case, based on the membership and vagueness values, each feature value and the case-rule judgments are represented by different fuzzy sets. Therefore, the similarity measures of cases become equivalent to similarity of fuzzy sets, which can be defined as the distance between the two centers of gravity. Let the membership function of a fuzzy set A be μ_A. The center of gravity of A for $x \in [a_1, a_2]$ is then computed by [also refer to equation (3.15)]

$$\text{CG}(A) = \frac{\int_{a_1}^{a_2} x\mu_A(x)\, dx}{\int_{a_1}^{a_2} \mu_A(x)\, dx} \tag{6.7}$$

The fuzzy similarity between two fuzzy sets A and B is defined as [also refer to equation (3.16)]

$$SM(A, B) = (1 - |CG(A) - CG(B)|) \qquad (6.8)$$

The similarity of fuzzy frames, including several features for a given issue, is computed, based on equations (6.7) and (6.8), by taking the minimum similarity values of fuzzy sets, which represent different features.

According to the similarity measure, the solution(s) of the most similar case or cases can be selected as the proposed solution(s) for the query case, and then an inference about whether the retrieved precedent conclusion can be adapted is made by the system using the case rules. The similarity measure defined in equation (6.8) is performed on the fuzzy sets representing the judgment of case rule for the precedent and query case. If the degree of similarity is greater than a given threshold, the query case conclusion can be obtained according to that of the precedent.

Another system, called HILDA [18], which uses ANNs for knowledge extraction, has been applied in the area of law. It incorporates rule- and case-based reasoning to assist a user in predicting case outcomes and generating arguments and case decisions. The knowledge extracted from an ANN guides the rule inferences and case retrieval.

6.6   PROPERTY VALUATION

The task of residential property valuation is to estimate the dollar value of properties in the dynamic real world. The most common and successful method used by expert appraisers is referred to as the *sales comparison approach*. The method consists of finding comparables (i.e., recent sales that are similar to the subject property using sale records); after contrasting the subject property with the comparables, adjusting the comparables' sales price to reflect their differences from the subject property; and reconciling the comparables' adjusted sales prices to derive an estimate for the subject property (using any reasonable averaging method). A system, Property Financial Information Technology (PROFIT), was developed by Bonissone and Cheetham [19] to automate this process. It uses fuzzy logic in case matching and retrieval to describe the user preference for case selection. It has been tested successfully on thousands of real estate transactions.

6.6.1   PROFIT System

PROFIT uses CBR with fuzzy predicates and fuzzy similarity measures [20] to estimate residential property value for real estate transactions. It consists of the following steps:

Step 1. Retrieve recent sales from a case base. Recent sales are retrieved from a case base using a small number of features to select potential comparables.

Step 2. Compare the subject property with the cases retrieved. The comparables are rated and ranked on a similarity scale to identify those most similar to the subject property. This rating is obtained from a weighted aggregation of the decision-maker preferences, expressed as fuzzy membership distributions and relations.

Step 3. Adjust the sale price of the cases retrieved. Each property's sale price is adjusted to reflect the cases differences from the subject property. These adjustments are performed by a rule set that uses additional property attributes, such as construction quality, conditions, pools, and fireplaces.

Step 4. Aggregate the adjusted sales prices of the cases retrieved. The best four to eight comparables are selected. The adjusted sale price and similarity of the properties selected are combined to produce an estimate of the subject value with an associated reliability value.

Step 2, in which fuzzy logic is incorporated to represent user preferences, is discussed in detail in the following section.

6.6.2 Fuzzy Preference in Case Retrieval

A set of potential comparables is extracted initially using standard SQL (structured query language) queries for efficiency. Retrieval involves comparing a number of specific attributes of the subject with those of each comparable. Then the concept of fuzzy preference is introduced to reflect user preferences, which greatly influence the similarities between the properties of the subject and comparables. These similarity values determine the rating or ranking of the comparables retrieved, and guide the selection and aggregation processes, leading to estimation of the final property value.

Six attributes—address, data of sale, living area, lot area, number of bathrooms, and number of bedrooms—are used in the initial case selection because their values are not missing in over 95% of the records in the database. Users' preference for the

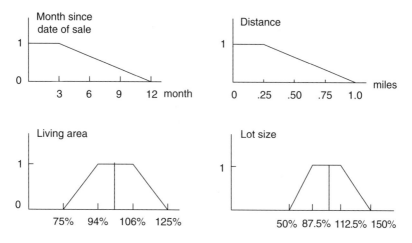

Figure 6.6 Attribute preference functions.

TABLE 6.7   Preference Function for Number of Bedrooms

Subject	Comparable					
	1	2	3	4	5	6+
1	1.00	0.50	0.05	0.00	0.00	0.00
2	0.20	1.00	0.50	0.05	0.00	0.00
3	0.05	0.30	1.00	0.60	0.05	0.00
4	0.00	0.05	0.50	1.00	0.60	0.20
5	0.00	0.00	0.05	0.60	1.00	0.80
6+	0.00	0.00	0.00	0.20	0.80	1.00

first four attributes can be expressed as trapezoidal fuzzy membership functions (shown in Fig. 6.6). For each attribute, the support set of the preference function represents the range of tolerable values; and the core represents the most desirable range of the top preference with the membership value 1. The remaining two features, number of bathrooms and number of bedrooms, are evaluated in a similar way. For example, the preference function of the number of bedrooms is represented by reflexive asymmetric fuzzy relation, which is illustrated in Table 6.7. As a result of the evaluation, each comparison gets a preference vector, whose value lies in the interval [0,1]. These values represent the partial degrees of membership of each feature value in the fuzzy sets and fuzzy relations, which represented users' selection preference.

The similarity measure is a function of the preference vectors and of the users' priorities. These priorities are represented using the weights reflecting the attributes' relative importance in the specific market area. For example, in Bonissone's experiment, the weights that are listed in the weight column can be obtained by interviewing expert appraisers using Saaty's pair wise comparison methods [21]. The evaluation representing the attributes' degree of matching obtained using the above membership functions, are listed in the preference column. The product of the preference score and the weights for an attribute is given in the weighted preference column. The similarity measure is then obtained as in Table 6.8. With the

TABLE 6.8   Similarity Measure Computation

Attribute	Subject	Comparable	Comparision	Preference	Weight	Weighted Preference
Months since data of sale	×	6 months	6 months	0.67	0.222	0.1489
Distance	×	0.2 mile	0.2 mile	1.00	0.222	0.2222
Living area	2,000	1,800	90%	0.79	0.333	0.2633
Lot size	20,000	35,000	175%	0.75	0.111	0.0367
No. bedrooms	3		0%	1.00	0.056	0.0556
No. bathrooms	2.5	2	2.5 > 2	0.75	0.056	0.0417
Similarity measure (sum of weighted preference/sum of weights)						0.768333

similarity measure, the initially extracted potential comparables are ranked and then one or more candidates are selected from them.

6.7 CORPORATE BOND RATING

We describe here the hybrid approach of Shin and Han [22] using genetic algorithms (GAs) to support CBR for corporate bond rating. GA is applied to find an optimal or nearly optimal weight vector for the attributes of cases in case indexing and retrieving. These weight vectors are used in the matching and ranking procedures of CBR, which guide effective retrievals of useful cases.

6.7.1 Structure of a Hybrid CBR System Using GAs

The overall structure of the hybrid approach is shown in Figure 6.7. It has three phases:

Phase 1. Search an optimal or nearly optimal weight vector with precedent cases (or reference cases) using GAs.

Phase 2. Apply the derived weight vector obtained from phase 1 to the case indexing scheme for the case-based retrieval process and evaluate the resulting model with additional validation cases (or test cases) having known outcomes.

Phase 3. Present new (unclassified) data to the model for solving.

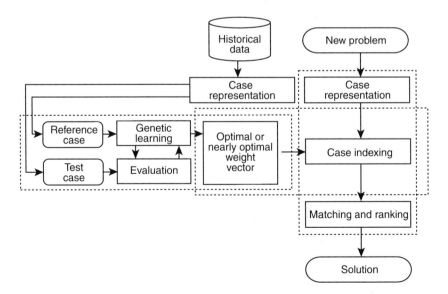

Figure 6.7 Structure of a hybrid GA–CBR system. (From [22].)

6.7.2 GA in Case Indexing and Retrieval

In phase 1, GA is introduced to determine optimal weights from the historical cases. The weight vectors assign importance values to each feature and are used in computing the similarity between the cases (for the purpose of case retrieval). The algorithm involves primarily the application of crossover and mutation operators to generate a new population for the problem solution, obtaining good solution(s) by evaluating the fitness of different weight vectors. The process is illustrated in Figure 6.8.

Depending on the problem, the weight values are encoded in a string, called a *chromosome*, and the range of the weights is set in [0,1]. The classification accuracy rate of the test set is used as the fitness function, which is expressed as

$$CR = \frac{1}{n}\sum_{i=1}^{n} CA_i \tag{6.9}$$

such that

$$CA_i = \begin{cases} 1 & \text{if } O(T_i) = O(S_{j^*(i)}) \\ 0 & \text{otherwise,} \end{cases}$$

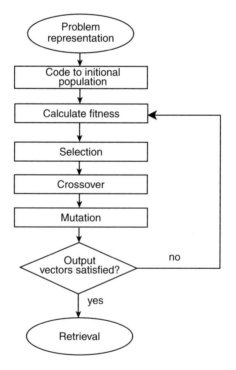

Figure 6.8   Genetic algorithm for determining optimal weight values for case retrieval.

where

$$S_{j^*(i)} = \min_{j \in R} \sqrt{\sum_{k=1}^{l} w_k (T_{ik} - R_{jk})^2} \qquad (6.10)$$

CR is the overall classification accuracy rate of the test set; CA_i is the classification accuracy of the ith case of the test set denoted by 1 and 0 (1 means correct and 0 means incorrect); $O(T_i)$ is the target output of ith case of the test set; $O(S_{j^*(i)})$ is the output of jth case of the reference set that has the minimum distance with the ith case of the test set; $S_{j*(i)}$ is the distance between the ith case of the test set and the jth case in the reference set R; T_{ik} is the kth feature of the ith case of the test set (T); R_{jk} is the kth feature of the jth case in the reference set; w_k is the importance (or weight) of the kth feature of a case; l denotes the number of features; and n is the number of test cases. As a stopping criterion, 2500 trials were used.

The experiment was performed on data consisting of 168 financial ratios (i.e., features) and corresponding bond ratings of Korean companies. The ratings had been performed by the National Information and Credit Evaluation Inc., which is one of the most prominent bond-rating agencies in Korea. The total number of samples available from 1991 to 1995 includes 3886 companies. The method was found experimentally to support effective retrieval of cases and to increase the overall classification accuracy rate significantly.

6.8 COLOR MATCHING

General Electric Plastics (GEP), one of the world's largest plastics producers, currently provides a color-matching service to customers. Customers give a physical sample of the color plastic they want and GEP either finds a close match from their color libraries or formulates a new color to meet the customers' need. The methodology of CBR is incorporated here to reduce the cost and shorten the turnaround time. Fuzzy logic is applied in the frame of CBR to achieve a consistent measure of multiple criteria for case selection.

6.8.1 Structure of the Color-Matching Process

The color-matching process developed by Cheetham and Graf [23] is shown in Figure 6.9. It involves selecting colorants and their respective amounts to generate a color that satisfies the customer requirement when combined with the plastic. The color matcher places the physical color standard in the spectrophotometer and reads the spectrum of the color standard into the color-matching system. Next, the color matcher enters key information, such as the resin and grade of material to generate the match. Then the system searches its case base of previous matches for the "best" one(s) and adjusts them to produce a match for the new standard. If this new match is acceptable, the adapted loadings are saved in the database and the

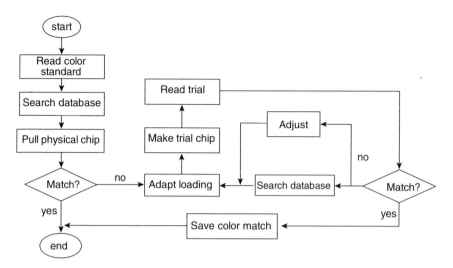

Figure 6.9 Color-matching process. (From [23].)

match is finished. If the new match is not acceptable, the system adapts it to match the requested color and application more closely. The color matcher then makes a physical chip using the adapted formula until a match is found. When the "end" oval is reached, a formula is obtained that gives the "best" color match and balance of all other important properties.

6.8.2 Fuzzy Case Retrieval

There are multiple criteria that the color match must satisfy: for example, the color of the plastic must match the standard under multiple lighting conditions, there must be enough pigments to hide the color of the plastic, and the cost of colorant formula should be as low as possible. The principle of nearest neighbor is applied in case retrieval. The case selection needs to provide a consistent meaning of similarity for each attribute mentioned above to find the nearest neighbor. The consistency

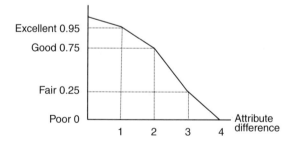

Figure 6.10 Example of the fuzzy preference function.

TABLE 6.9   Linguistic Terms and Similarity

Fuzzy Rating	Maximum Score	Minimum Score
Excellent	1	0.95
Good	0.94	0.75
Average/fair	0.74	0.25
Poor	0.24	0

is achieved through the use of fuzzy linguistic terms associated with measured differences, such as excellent, good, fair, and poor, to represent each attribute.

A fuzzy preference function is used to calculate the similarity of an attribute of a case with the corresponding attribute of the subject. For example, a difference of 1 unit in the values of that attribute for the subject and comparable would be considered as excellent, 2 would be good, 3 would be fair, and 4 would be poor. This rating is then transformed into the fuzzy preference function shown in Figure 6.10. The fuzzy preference function, which is used to transform a quantifiable value for each attribute into a qualitative description of the attribute, allows a comparison of properties that are based on entirely different scales, such as cost measured in cents per pound and spectral curve match measured in reflection units. Based on a discussion with experts and the classification analysis in terms of linguistic terms, it is found that there is enough precision in the evaluation of the similarity of the attributes to have four linguistic terms. Table 6.9 shows the linguistic terms and the similarity scores that correspond to them.

Fuzzy preference functions are created for each of the following attributes of the color match: color similarity, total colorant load, cost of colorant formula, optical density of color, and color shift when molded under normal and abusive conditions. As a result, a vector called the *fuzzy preference vector* is generated which contains a fuzzy preference value for each attribute. A summation of the preference value of each attribute can then be obtained with a weight of unity for each attribute, or with different weight values for different attributes if the end users desire to emphasis one attribute over another. Based on the summation of the preference value of each attribute, the similar cases can be retrieved from the case base, and the similarity calculation can also be used to guide the adaptation process.

6.9   SHOE DESIGN

A CBR system for fashion shoe design [24] is described in this section. Fuzzy sets are used to represent cases. Multilevel supervised neural nets are incorporated to carry out the task of case retrieval. The use of neural networks and fuzzy logic has been found to be a useful means to improve the retrieval accuracy. During testing, the cases retrieved are found to be the closest match of the cases in the case base in 95% of tests carried out. In the other 5%, the retrieved cases are still useful for adaptation, although not the closest possible match. The principal features of fuzzy representation and neural network–based retrieval methods are described here in brief.

6.9.1 Feature Representation

In shoe design, the features that define a case and index that case in a case base are of different types (i.e., Boolean, continuous, multivalued, and fuzzy), which need to be represented in different ways. In the following discussion, the representation of a fuzzy feature is described.

For example, consider a question: How high is the shoe? Is the footwear item a sandal, shoe, short boot, or long boot? It is explained by the example shown in Figure 6.11. The height of a shoe can be small (sandal), medium (shoe), high (short boot), or very high (long boot). The actual data are classified into these four ranges of values (categories). This is a problem, as values can be considered to be between groups or in more than one group. A pair with an inside back height of 30 cm would be considered very high (a long boot), and a pair with an inside back height of 5 cm would be considered medium (a shoe) and 10 cm high (a short boot). When it comes to a value of 7 cm, is that average or high? What about 8 or 9 cm? What is the boundary between a shoe and a short boot? A large number of the features that characterize shoe design cases frequently consist of linguistic variables that are best represented using fuzzy feature vectors.

The inside-back height of a shoe is classified as very high (a long boot), high (a short boot), medium (a shoe), or low (a type of sandal), but some shoes will not fit into the crisp separations of categories. This is represented by using four linguistic variables: "low," "medium," "high," and "very high." Each shoe would have a value for each of the four variables.

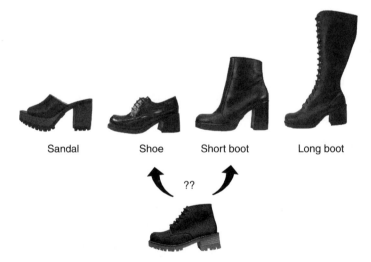

Sandal   Shoe   Short boot   Long boot

??

Figure 6.11 Footwear item between height categories.

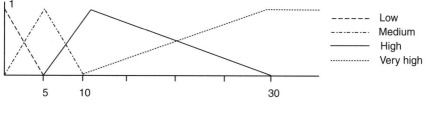

Height (cm)

Figure 6.12   Fuzzy representation of shoe height.

- A low shoe (e.g., a mule) would have attributes with these values: low, 1; medium, 0; high, 0; very high, 0.
- A long boot with a large inside-back height would have the following values: low, 0; medium, 0; high, 0; very high, 1.
- A traditional shoe (of 5 cm) would have the following values: low, 0; medium, 1; high, 0; very high, 0.

This is straightforward for the instances that fall exactly into one of the main categories. However, for those cases that do not fit so nicely (e.g., the 7-cm case), one needs to define a function that determines the value for in-between cases. Although this function can take many forms, the authors [24] used piecewise linear forms (Fig. 6.12) to determine the values associated with the height of the shoe.

6.9.2   Neural Networks in Retrieval

The shoes are divided into larger and smaller groups in a hierarchy; each case (i.e., a particular design) is attached to an appropriate subclass. For each classification, since a single-layer network is insufficient, a three-layered network (i.e., with one hidden layer) is used at each level for classification. Such a network builds its own internal representation of the inputs and determines the best weights. The back-propagation algorithm is used for its training.

Determination of features that are important for the classification of cases is a major knowledge engineering task. For example, to determine the features that should be used as inputs into the highest-level neural network, one needs to look at the classes into which the superclass is divided and how the cases in each of those subclasses differ from each other. Distinguishing the features between classes is not easy in the case of the shoe design problem. Determination of the correct features involves looking at the cases of previous designs, seeing the major differences between them, and using these features to train a network successfully. The method used is as follows: (1) when a network learns the set of input patterns to give a low error, the features are considered to be correct; (2) if there are patterns (or cases) that this network cannot classify, new features are added that can distinguish between these cases. Once the additional features are determined, the input patterns for the neural network are modified to include the additional attributes and the networks retrained. This training and testing cycle continued until all the networks had negligible error rates. After retrieval of the best case, the design is modified by making up the parts where the retrieved design does not meet the required design. Every time a new successful design was created, it was added to the case base.

To add a new case to the footwear design system, the case first needs to be classified or grouped as the best possible with similar designs. This may involve some change(s) to the hierarchical structure of the case base. Once the case has been classified, the neural networks used in its classification need retraining. This is simple when new features are not needed for the classification. If the addition of a new case requires the addition of some new input to a neural network, a new neural network has to be created and trained.

6.10 OTHER APPLICATIONS

A software system using CBR and fuzzy logic, named CREW [25], was developed to simulate a time-series model of astronaut crews. Without the expense and risk of actual flights, it can predict how a given crew might respond to mission stress under different scenarios. In the frame of CBR, some adaptation rules, coming from the statements of researchers, act as a supplement to the case base. The conditions of these rules are often described by linguistic terms such as "low," "medium," and "high" with modifiers such as "very" or "not." The nonquantitative nature of these rules lends itself to formulation and computation using fuzzy logic. They are translated reasonably into numeric computations by the defined membership functions, preserving sufficient precision considering the limited effect that any given rule or condition has on the overall simulation.

In 2002, Passone et al. [26] provided an application of CBR technology to support the design of estuarine models (i.e., models for evaluating water quality). The system aims to help nonexpert users to select a model that matches their goal and the nature of the problem to be solved. It consists of three components: a case-based reasoning scheme, a genetic algorithm, and a library of numerical estuarine models. Once an appropriate model procedure is selected using the case-based reasoning scheme, a genetic algorithm with problem-specific knowledge is activated to adjust the model parameters. An example based on the Upper Milford Haven estuary in the United Kingdom is used to demonstrate the efficiency of the system's structure for supporting estuarine model design.

6.11 SUMMARY

In this chapter we have briefly described some soft CBR applications so that the reader gains some insight into the scope of these application areas. The soft CBR systems show improvement of case retrieval accuracy, handling of multiple selection criteria, constructing a user's preference membership function, as well as improvement in adaptation ability. Especially, use of the membership function allows a greater accuracy and smoothness in the selection phase. Handling of multiple selection criteria enables the system to detect the potential problems during the selection step, and incorporating the generalization ability of ANN enhances the performance of case adaptation.

REFERENCES

1. C. K. P. Wong, S. C. K. Shiu, and S. K. Pal, Mining fuzzy association rules for Web access case adaptation, in *Proceedings of Soft Computing in Case-Based Reasoning Workshop, in Conjunction with the Fourth International Conference in Case-Based Reasoning (ICCBR-01)*, Vancouver, British Columbia, Canada, Springer-Verlag, Berlin, pp. 213–220, 2001.

2. C. K. P. Wong, Web access path prediction using fuzzy case-based reasoning, M.Phl. thesis, Department of Computing, Hong Kong Polytechnic University, Hong Kong, 2003.

3. C. C. Hsu and C. S. Ho, A hybrid case-based medical diagnosis system, in *Proceedings of the Tenth IEEE International Conference on Tools with Artificial Intelligence (ICTAI-98)*, Taipei, Taiwan, pp. 359–366, 1998.

4. C. C. Hsu and C. S. Ho, Acquiring patient data by an intelligent interface agent with medicine-related common sense reasoning, *Expert Systems with Applications: An International Journal*, vol. 17, no. 4, pp. 257–274, 1999.

5. R. S. Cotran, V. Kumar, and S. L. Robbins, *Robbins Pathology Basis of Disease*, W. B. Saunders, Philadelphia, 1989.

6. A. S. Fauui, E. Brauwald, K. J. Isselbacker, J. D. Wilson, J. B. Martin, D. L. Kasper, S. L. Hauser, and D. L. Longo, *Harrison's Principles of Internal Medicine*, 14th ed., McGraw-Hill, New York, 1998.

7. A. Aamodt, Case-based reasoning: foundational issues, methodological variations, and system approaches, *Artificial Intelligence Communications*, vol. 7, no. 1, pp. 39–59, 1994.

8. P. Cuddihy and R. Shah, Method and system for analyzing error logs for diagnostics, U.S. patent 5,463,768, 1995.

9. J. M. Corchado and B. Lees, Adaptation of cases for case-based forecasting with neural network support, in *Soft Computing in Case Based Reasoning* (S. K. Pal, S. D. Tharam, and D. S. Yeung, eds.), Springer-Verlag, London, pp. 293–320, 2001.

10. J. M. Corchado, N. Rees, and J. Aiken, Internal report on supervised ANNs and oceanographic forecasting, PML, Plymouth, Devonshire, England, December 30, 1996.

11. C. R. Bishop, *Neural Networks for Pattern Recognition*, Clarendon Press, Oxford, 1995.

12. K. H. Bjarne and R. Denis, Weather prediction using case-based reasoning and fuzzy set theory, in *Proceedings of Soft Computing in Case-Based Reasoning Workshop, in Conjunction with the Fourth International Conference on Case-Based Reasoning (ICCBR-01)*, Vancouver, British Columbia, Canada, Springer-Verlag, Berlin, pp. 175–178, 2001.

13. J. M. Corchado, B. Lees, C. Fyfe, N. Rees, and J. Aiken, Neuro-adaptation method for a case-based reasoning system, in *Proceedings of the International Joint Conference on Neural Networks (IJCNN-98)*, Skövde, Sweden, IEEE Press, Piscataway, NJ, pp. 713–718, 1998.

14. J. M. Corchado, B. Lees, C. Fyfe, and N. Rees, Adaptive agents: learning from the past and forecasting the future, in *Proceedings of the First International Conference on the*

Practical Application of Knowledge Discovery and Data Mining (PADD-97), London, pp. 109–123, 1997.

15. K. Hirota, H. Yoshino, and M. Q. Xu, An application of fuzzy theory to the case-based reasoning of the CISG, *Journal of Advanced Computational Intelligence*, vol. 1, no. 2, pp. 86–93, 1997.

16. K. Hirota, H. Yoshino, M. Q. Xu, Y. Zhu, and D. Horie, A fuzzy case base reasoning system for the legal inference, in *Proceedings of Fuzzy Systems, World Congress on Computational Intelligence*, Anchorage, AK, pp. 1350–1354, 1998.

17. M. Q. Xu, K. Hirota, and H. Yoshino, A fuzzy theoretical approach to representation and inference of case in CISG, *International Journal of Artificial Intelligence and Law*, vol. 7, no. 2–3, pp. 259–272, 1999.

18. P. A. Egri and P. F. Underwood, HILDA: knowledge extraction from neural networks in legal rule based and case base reasoning, in *Proceedings of the IEEE International Conference on Neural Networks (ICNN-95)*, Perth, Western Australia, vol. 4, pp. 1800–1805, 1995.

19. P. P. Bonissone and W. Cheetham, Financial application of fuzzy case-based reasoning to residential property valuation, in *Proceedings of the Sixth IEEE International Conference on Fuzzy Systems (FUZZ-IEEE-97)*, Barcelona, pp. 37–44, 1997.

20. P. P. Bonissone and S. Ayub, Similarity measures for case-based reasoning systems, in *Proceedings of the Fourth International Conference on Information Processing and Management of Uncertainty (IPMU-92) in Knowledge-Based Systems*, Palma de Mallorca, Spain, pp. 483–487, 1992.

21. T. L. Saaty, *The Analytic Hierarchy Process*, McGraw-Hill, New York, 1980.

22. K. S. Shin and I. Han, Case-based reasoning supported by genetic algorithms for corporate bond rating, *Expert Systems with Applications*, vol. 16, pp. 85–95, 1999.

23. W. Cheetham and J. Graf, Case-based reasoning in color matching, in *Proceedings of the Second International Conference on Case-Based Reasoning (ICCBR-97)*, Providence, RI, Springer-Verlag, Berlin, vol. 1266, pp. 1–12, 1997.

24. J. Main and T. S. Dillion, Hybrid case based reasons for footwear design, in *Proceedings of the Third International Conference on Case-Based Reasoning (ICCBR-99)*, Seeon Monastery, Munich, Springer-Verlag, Berlin, pp. 499–509, 1999.

25. G. Stahl, Armchair mission to Mars: using case-based reasoning and fuzzy logic to simulate a time series model of astronaut crews, in *Soft Computing in Case Based Reasoning* (S. K. Pal, T. S. Dillon, and D. S. Yeung, eds.), Springer-Verlag, London, pp. 321–334, 2001.

26. S. Passone, W. H. Chung, and V. Nassehi, Case-based reasoning for estuarine model design, in *Proceedings of the Sixth European Conference on Case-Based Reasoning (ECCBR-02)*, Aberdeen, Scotland, pp. 590–603, 2002.

APPENDIXES

In the appendixes, the basic concepts, definitions, and characteristic features of four soft computing techniques—fuzzy logic, artificial neural networks, genetic algorithms, and rough sets—are described. The treatment of these topics here is both introductory and fundamental. Readers will find this background knowledge useful for an understanding of the ideas and techniques presented in the text.

In Appendix A we define fuzzy subsets, membership functions, basic operations, measure of fuzziness, and fuzzy classification. In Appendix B, the architecture, training methods, and advantages of artificial neural networks are described. Here, four basic neural network models—perceptron, multilayer perceptron, radial basis function neural network, and Kohonen neural network—which are widely used, are considered. In Appendix C, the basic principle, algorithm, and merits of genetic algorithms are given. Finally, rough set theory is presented in Appendix D. This includes information system, indiscernibility relation, set approximations, rough membership, and the dependency between attributes.

Note that the tools above act synergistically, not competitively, for enhancing the problem-solving ability of each other. The purpose is to provide flexible information-processing systems that can exploit the tolerance for imprecision, uncertainty, approximate reasoning, and partial truth in order to achieve tractability, robustness, low solution cost, and close resemblance to human decision making.

Foundations of Soft Case-Based Reasoning.   By Sankar K. Pal and Simon C. K. Shiu
ISBN 0-471-08635-5   Copyright © 2004 John Wiley & Sons, Inc.

APPENDIX A

FUZZY LOGIC

Fuzzy logic (FL), initiated in 1965 by Lotfi Zadeh [1], is a powerful problem-solving technology. It provides a simple and useful way to draw definite conclusions from vague, ambiguous, or imprecise information. Basically, fuzzy logic is an extension of Boolean logic. In fuzzy logic, the truth value of a proposition lies between 0 and 1. This allows expressing the knowledge with subjective concepts such as tall, short, cold, fast, and slow. In a sense, fuzzy logic resembles human decision making in its use of approximate information and uncertainty to generate decisions. There are two important concepts within fuzzy logic that play a central role in its applications [2]:

- *Linguistic variable:* a variable whose values are words or sentences in a natural or synthetic language
- *Fuzzy IF–THEN rule:* a production rule in which the antecedent and consequent are propositions containing linguistic variables

The basic concepts of fuzzy logic are described below; interested readers may refer to references such as [1–19] for more detail information.

Foundations of Soft Case-Based Reasoning.   By Sankar K. Pal and Simon C. K. Shiu
ISBN 0-471-08635-5   Copyright © 2004 John Wiley & Sons, Inc.

A.1 FUZZY SUBSETS

A crisp subset A of U (i.e., universe of discourse) is defined by a characteristic function χ_A, which takes a value 1 for the elements that belong to A and zero for those elements that do not belong to A.

$$\chi_A(x): \ U \rightarrow \{0, 1\}$$

$$\chi_A(x) = \begin{cases} 0 & \text{if } x \notin A \\ 1 & \text{if } x \in A \end{cases}$$

A fuzzy subset A of U is defined by a membership function that associates to each element x of U a membership value $\mu_A(x)$. This value represents the grade of membership (i.e., between 0 and 1) of x to A:

$$\mu_A(x): \ U \rightarrow [0, 1]$$

Note that if μ_A only takes value 0 or 1, the fuzzy subset A becomes a crisp subset of U. Thus, a crisp subset can be regarded as a special case of a fuzzy subset.

In general, a fuzzy subset A is represented as follows:

- If U is finite [i.e., $U = \{x_1, x_2, \ldots, x_n\}$], fuzzy subset A can be denoted as

$$A = \{(\mu_A(x_i),\ x_i),\ i = 1, 2, \ldots, n\} = \left\{ \frac{\mu_A(x_i)}{x_i},\ i = 1, 2, \ldots, n \right\} \tag{A.1a}$$

$$A = \sum_{x \in U} \frac{\mu_A(x)}{x} = \frac{\mu_A(x_1)}{x_1} + \frac{\mu_A(x_2)}{x_2} + \cdots + \frac{\mu_A(x_n)}{x_n} \tag{A.1b}$$

where "$+$" denotes union.

- If U is infinite [i.e., $U = \{x_1, x_2, \ldots, x_n, \ldots\}$],

$$A = \int \frac{\mu_A(x)}{x}$$

In the following, some concepts of fuzzy subset A are given.

- *Support.* The support of A, denoted as support(A), is the crisp set of all points in U such that the membership function of A is nonzero:

$$\text{support}(A) = \{x \mid x \in U, \mu_A(x) > 0\}$$

Note: If the support of a fuzzy set contains only one single element $x_1 \in U$, then $A = \mu_1/x_1$ is called a *fuzzy singleton*. If $\mu_1 = 1$, then $A = 1/x_1$ is called a *nonfuzzy singleton.*

- *Kernel.* All the elements that belong absolutely (i.e., with degree 1) to A constitute the kernel of A, denoted as $\ker(A)$:

$$\ker(A) = \{x | x \in U, \mu_A(x) = 1\}$$

If A is a crisp subset of U, it is called *normalized*. In this case, A is identical to its support and its kernel.

- *Cardinality.* This is the sum of the membership degrees of all the elements of U belonging to A:

$$|A| = \sum_{x \in U} \mu_A(x)$$

If A is a crisp subset of U, its cardinality represents the number of elements of A.

An example that describes a set of young people using a fuzzy subset is given below. In general, if a person is less than 20 years old, he or she will be considered as "young." The fuzzy subset of "young" therefore can be characterized by the following equation:

$$\mu_{young} = \begin{cases} 1 & x < 20 \\ -\frac{1}{20}x + 2 & 20 \le x < 40 \\ 0 & x \ge 40 \end{cases}$$

where x is a element of $U = \{0, 1, 2, \ldots, 50\}$. Let this fuzzy subset be denoted as A, and thus

$$\text{support}(A) = \{0, 1, \ldots, 39\}, \ker(A) = \{0, 1, \ldots, 20\}, \quad \text{and} \quad |A| = 30.5$$

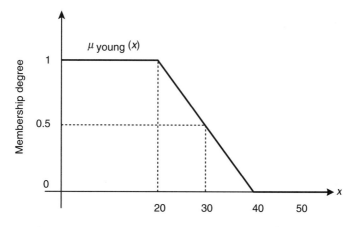

Figure A.1   Membership function of fuzzy subset "young."

From Figure A.1, people in the age group [0,20] (i.e., $0 \le x \le 20$) are absolutely members of the set "young" (i.e., they have a membership degree of 1). People in the group [20,30] are partial members of the set "young" (e.g., a person 30 years old is "young" with a membership degree of 0.5). People over 40 are nonmembers of the set "young." Note that there are many possible membership functions that can be used to characterize the fuzzy subset "young."

- *Linguistic variable.* A linguistic variable is a variable whose values are words (i.e., linguistic terms) or sentences in a natural or synthetic language. For example, for a linguistic variable such as "age," its values can be "very young," "young," "middle-aged," "old," "very old," and so on. A linguistic variable can be interpreted as a label of a fuzzy set that is characterized by a membership function, and its formal definition is given below.

Definition A.1. A *linguistic variable* is characterized by the quintuple $\langle X, T, U, g, s \rangle$, in which X is the name of the variable whose values range over the universal set U, T is a finite set of linguistic terms $\{t_0, t_1, \ldots, t_n\}$ that describes X, g a grammar for generating linguistic terms, and s a semantic rule mapping each term $t \in T$ to a fuzzy subset [i.e., $s(t)$] on U.

For example, for describing the age of the students in a university, the linguistic variable "age" can be used. From Definition A.1, X = age and U consists of the possible values of the age of a university student [e.g., $U = \{0, 1, 2, \ldots, 90\}$]. These values are described by a preordered set T, such as $T = \{\text{young, middle-aged, old}\}$. The grammar g is used for generating additional terms, such as "very young" and "very old." The semantic rule s maps each term, such as "young," to a fuzzy subset [i.e., $s(\text{young})$].

A.2   MEMBERSHIP FUNCTIONS

A membership function specifies the degree of membership of an element to a fuzzy set. Determination of membership functions is subjective in nature and context dependent. Some popular membership functions are triangular membership function, trapezoid membership function, sigmoid membership function, Gaussian membership function, bell membership function, S function, and π function. The simpler ones are described below.

- *Triangle membership function.* This membership function (see Fig. A.2) is defined as

$$\text{triangle}(x; a, b, c) = \max\left\{\min\left\{\frac{x-a}{b-a}, \frac{c-x}{c-b}\right\}, 0\right\} \qquad (A.2)$$

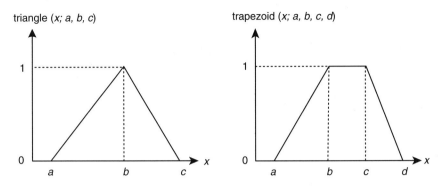

Figure A.2   Triangle membership function.   **Figure A.3**   Trapezoid membership function.

- *Trapezoid membership function.* This membership function (see Fig. A.3) is defined as

$$\text{trapezoid}(x; a, b, c, d) = \max\left\{\min\left\{\frac{x-a}{b-a}, \ 1, \ \frac{d-x}{d-c}\right\}, \ 0\right\} \tag{A.3}$$

Note that the triangle membership function is a special form of the trapezoid membership function. These straight-line membership functions have the advantage of simplicity.

- *Sigmoid membership function.* This membership function (see Figs. A.4 and A.5) is defined as

$$\text{sigmoid}(x; a, c) = \frac{1}{1 + e^{-a(x-c)}} \tag{A.4}$$

From Figures A.4 and A.5, if the parameter $a > 0$, this function is continuous and open on the right side, while if $a < 0$, this function is continuous and open on the left side. This function can be used for describing concepts such as "large" or "small" and are often used in neural networks as an activation function.

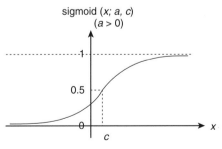
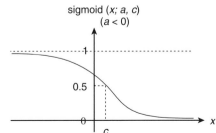

Figure A.4   Sigmoid membership function   **Figure A.5**   Sigmoid membership function
$(a > 0)$. $(a < 0)$.

Although the membership functions are subjective, they cannot be defined arbitrarily. For example, a fuzzy set of "integers around b" can be represented by some function as in Figure A.2, in which the membership value first increases monotonically until $x = b$, where it attains a maximum value of unity, and then decreases; and therefore this fuzzy set can not be represented by a function as in Figure A.4 or A.5.

A.3   OPERATIONS ON FUZZY SUBSETS

Basic operations related to fuzzy subsets A and B of U having membership degrees $\mu_A(x)$ and $\mu_B(x)$, $x \in U$, respectively, are given below:

- A is equal to $B(A = B) \Rightarrow \mu_A(x) = \mu_B(x)$ for all $x \in U$.
- A is a complement of $B(A = \bar{B}) \Rightarrow \mu_A(x) = \mu_{\bar{B}}(x) = 1 - \mu_B(x)$ for all $x \in U$.
- A is contained in $B(A \subseteq B) \Rightarrow \mu_A(x) \leq \mu_B(x)$ for all $x \in U$.
- The union of A and $B(A \cup B) \Rightarrow \mu_{A \cup B}(x) = \vee(\mu_A(x), \mu_B(x))$ for all $x \in U$, where $\vee$ denotes maximum.
- The intersection of A and $B(A \cap B) \rightarrow \mu_{A \cap B}(x) = \wedge(\mu_A(x), \mu_B(x))$ for all $x \in U$, where $\wedge$ denotes minimum.
- The concentration (Con) and dilation (Dil) operators for a fuzzy subset A are characterized, respectively, by

$$\mu_{\text{Con}(A)}(x) = (\mu_A(x))^2$$
$$\mu_{\text{Dil}(A)}(x) = (\mu_A(x))^{1/2}$$
(A.5)

- The modifiers, such as "not," "very," and "more or less" on fuzzy subset A can be characterized as

$$\mu_{\text{not } A} = 1 - \mu_A \tag{A.6}$$
$$\mu_{\text{very } A} = (\mu_A)^2 \tag{A.7}$$
$$\mu_{\text{not very } A} = 1 - (\mu_A)^2 \tag{A.8}$$
$$\mu_{\text{more or less } A} = (\mu_A)^{0.5} \tag{A.9}$$

A.4   MEASURE OF FUZZINESS

The measure of fuzziness, denoted as FM, of a fuzzy set indicates on a global level the average amount of difficulty in determining whether or not an element belongs to a set. In general, it should have the following properties:

- $FM(A) = \begin{cases} \text{minimum} & \text{iff} \quad \forall\, x_i \quad \mu_A(x_i) = 0 \text{ or } 1 \\ \text{maximum} & \text{iff} \quad \forall\, x_i \quad \mu_A(x_i) = 0.5 \end{cases}$

- $FM(A) \geq FM(A^*)$, where A^* is a *sharpened* version of A, defined as

$$\begin{aligned} \mu_{A^*}(x_i) \geq \mu_A(x_i) & \qquad \text{if } \mu_A(x_i) \geq 0.5 \\ \mu_{A^*}(x_i) \leq \mu_A(x_i) & \qquad \text{if } \mu_A(x_i) \leq 0.5 \end{aligned}$$

- $FM(A) = FM(\overline{A})$, where $\overline{A}$ is the complement set of A.

Different methods for computing FM have been proposed by [9,10,20–23]. One of the most popular measures is given here [10]: The FM of a fuzzy set A having n supporting elements can be computed as

$$FM(A) = \frac{2}{n^k} d(A, \underline{A}) \tag{A.10}$$

where $d(A, \underline{A})$ denotes the distance between A and its nearest crisp set $\underline{A}$, which is defined by

$$\mu_{\underline{A}}(x) = \begin{cases} 0 & \text{if} \quad \mu_A(x) \leq 0.5 \\ 1 & \text{if} \quad \mu_A(x) > 0.5 \end{cases} \tag{A.11}$$

Note that, the value of k depends on the distance metric being used. For example, if d is Hamming distance, $k = 1$, and

$$FM(A) = \frac{2}{n} \sum_{i=1}^{n} |\mu_A(x_i) - \mu_{\underline{A}}(x_i)| \tag{A.12}$$

For Euclidean distance, $k = \frac{1}{2}$, and

$$FM(A) = \frac{2}{\sqrt{n}} \sqrt{\sum_{i=1}^{n} \left[\mu_A(x_i) - \mu_{\underline{A}}(x_i)\right]^2} \tag{A.13}$$

A.5 FUZZY RULES

In conventional (crisp) classification problem, a sample is considered either a member of a class or not a member (i.e., the degree of class membership is binary). In fuzzy classification, a sample is considered to belong to more than one class (i.e., different classes) with different degrees of membership. The task of fuzzy classification is to select the best-fit class to which the particular sample in question belongs. In the following sections, an example of fuzzy classification is presented.

A.5.1 Definition

Linguistic variables and fuzzy sets are used as the antecedents and consequences of fuzzy rules. These rules are also called fuzzy IF–THEN rules. The definition of a fuzzy rule is given in Definition A.2.

Definition A.2. A fuzzy IF–THEN rule relates m antecedent variables, A_1, $A_2, \ldots, A_m$, to n consequent variables, $B_1, B_2, \ldots, B_n$, and has the form

$$\text{IF } [X_1 = A_1] \text{ AND } [X_2 = A_2], \ldots, \text{AND } [X_m = A_m] \text{ THEN}$$
$$[Y_1 = B_1] \text{ AND } [Y_2 = B_2], \ldots, \text{AND } [Y_n = B_n]$$

where $X = (X_1, X_2, \ldots, X_m)$ and $Y = (Y_1, Y_2, \ldots, Y_n)$ are linguistic variables, and $(A_1, A_2, \ldots, A_m)$ and $(B_1, B_2, \ldots, B_n)$ are their corresponding linguistic values. Here all linguistic values are in the form of fuzzy sets with membership functions such as μ_{A_i} and μ_{B_i}.

An example of a fuzzy IF–THEN rule is

$$\text{IF[pressure} = \text{``low''] AND [temperature} = \text{``high''] THEN [volume} = \text{``big'']}$$

A collection of fuzzy rules constitutes a fuzzy rule base that can be used to provide advice or decision support to input queries.

A.5.2 Fuzzy Rules for Classification

Consider the following fuzzy rule:

$$\text{IF [feature } F_1 = \text{``low''] AND [feature } F_2 = \text{``medium'']}$$
$$\text{AND [feature } F_3 = \text{``medium''] AND [feature } F_4 = \text{``medium'']}$$
$$\text{THEN [class} = \text{``class 4'']}$$

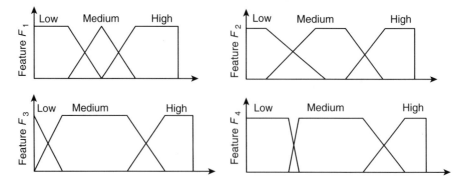

Figure A.6 Membership functions of the linguistic terms in fuzzy rule.

TABLE A.1   Example for a Fuzzy Rule Base

Rule	Feature				Class
	F_1	F_2	F_3	F_4	
r_1	Low	Medium	Medium	Medium	1
r_2	Medium	High	Medium	Low	2
r_3	Low	High	Medium	High	3
r_4	Low	High	Medium	High	1
r_5	Medium	Medium	Medium	Medium	4
...	...	...	...	...	...
r_n	Low	High	Medium	Low	5

where each linguistic term, such as "low" and "medium," is described by fuzzy sets shown in Figure A.6. A fuzzy rule base, consisting of n rules, is shown in Table A.1. For example, r_1: **IF** $[F_1 = $ "low"$]$ **AND** $[F_2 = $ "medium"$]$ **AND** $[F_3 = $ "medium"$]$ **AND** $[F_4 = $ "medium"$]$ **THEN** [Class $=$ class 1] is the first fuzzy rule in the rule base.

Given an input $x = (x_1, x_2, x_3, x_4)$, the procedure to determine its class label is as follows:

Step 1. Compute the membership degrees for each element of x (i.e., x_1, x_2, x_3, x_4) with respect to all the fuzzy rules. For example, for r_1 the degree of membership of x_1 to low is 0.75; the degree of membership of x_2 to medium, x_3 to medium, and x_4 to medium is 0.32, 0.6, and 0.5, respectively.

Step 2. Determine the overall truth degree of each rule by taking the minimum of membership degrees of all the antecedents. That is, in r_1, the truth degree is 0.32, as computed by

$$\begin{aligned} \text{degree}(r_1) &= \min\{\mu_{\text{low}}(x_1), \mu_{\text{medium}}(x_2), \mu_{\text{medium}}(x_3), \mu_{\text{medium}}(x_4)\} \\ &= \min\{0.75, 0.32, 0.6, 0.5\} \\ &= 0.32 \end{aligned}$$

Step 3. Determine the membership degree of x to each class (e.g., class 1) by taking the maximum of the membership degrees of all the rules where "class 1" is the consequence part. Assuming only two rules r_1 and r_4 are being considered, and both of them are having "class 1" as the consequence, and degree$(r_4) = 0.56$; thus,

$$\text{degree(class 1)} = \max\ \{\text{degree}(r_1), \text{degree}(r_4)\} = \max\{0.32, 0.56\} = 0.56$$

Therefore, a fuzzy set with respect to x is generated and represented, considering all the classes, as

$$A = \frac{\text{degree(class 1)}}{\text{class 1}} + \frac{\text{degree(class 2)}}{\text{class 2}} + \cdots + \frac{\text{degree(class 5)}}{\text{class 5}}$$

Step 4. To obtain a crisp classification, the fuzzy output (i.e., *A*) has to be defuzzified. There are several heuristic-based methods for defuzzification, such as the maximum method, where the class with the maximum degree to the input is chosen. That is, if

$$\text{degree(class } i) = \max\{\text{degree(class } 1), \text{degree (class } 2), \ldots, \text{degree (class } 5)\}$$

class i is chosen as the crisp class (to which the input belongs) with the crisp output value degree(class *i*).

Another widely used defuzzification method is to take the center of gravity of the fuzzy set as the crisp output value. For more details, readers may refer to [2,18].

REFERENCES

1. L. A. Zadeh, Fuzzy sets, *Information and Control*, vol. 8, pp. 338–353, 1965.
2. L. A. Zadeh, Soft computing and fuzzy logic, *IEEE Software*, vol. 11, no. 6, pp. 48–56, 1994.
3. M. Banerjee and S. K. Pal, Roughness of a fuzzy set, *Information Sciences*, vol. 93, no. 3–4, pp. 235–246, 1996.
4. J. C. Bezdek, *Pattern Recognition with Fuzzy Objective Function Algorithms*, Plenum Press, New York, 1981.
5. J. C. Bezdek and S. K. Pal (eds.), *Fuzzy Models for Pattern Recognition: Methods That Search for Structures in Data*, IEEE Press, Piscataway, NJ, 1992.
6. D. Dubois and H. Prade, *Fuzzy Sets and Systems: Theory and Applications*, Academic Press, San Diego, CA, 1980.
7. M. Hellmann, Fuzzy logic introduction, *http://epsilon.nought.de/*, 2001.
8. A. Kandel, *Fuzzy Techniques in Pattern Recognition*, Wiley-Interscience, New York, 1982.
9. A. Kandel, *Fuzzy Mathematical Techniques with Applications*, Addison-Wesley, Reading, MA, 1986.
10. A. Kaufmann, *Introduction to the Theory of Fuzzy Subsets: Fundamental Theoretical Elements*, Academic Press, San Diego, CA, 1975.
11. A. Kaufmann and M. Gupta, *Introduction to Fuzzy Mathematics*, Van Nostrand Reinhold, New York, 1985.
12. G. J. Klir and B. Yuan, *Fuzzy Sets and Fuzzy Logic: Theory and Applications*, Prentice Hall, Upper Saddle River, NJ, 1995.
13. S. K. Pal and D. Dutta Majumder, *Fuzzy Mathematical Approach to Pattern Recognition*, Wiley (Halsted), New York, 1986.
14. S. K. Pal and S. Mitra, *Neuro-Fuzzy Pattern Recognition: Methods in Soft Computing*, Wiley, New York, 1999.
15. L. A. Zadeh, Outline of a new approach to the analysis of complex systems and decision processes, *IEEE Transaction on Systems, Man, and Cybernetics*, vol. 3, pp. 28–44, 1973.
16. L. A. Zadeh, The concept of a linguistic variable and its application to approximate reasoning: parts 1, 2, and 3, *Information Sciences,* vols. 8, 8, and 9, pp. 199–249, 301–357, 43–80, 1975.

17. L. A. Zadeh, The role of fuzzy logic in the management of uncertainty in an expert system, *Fuzzy Sets and Systems*, vol. 11, pp. 199–227, 1983.

18. L. A. Zadeh, Fuzzy logic, *Computer*, vol. 21, pp. 83–92, 1988.

19. Aptronix, Inc., What is fuzzy logic, *Fuzzy Inference Development Environment (FIDE)*, *http://www.aptronix.com/fide/whatfuzzy.htm*, 2000.

20. B. Kosko, Fuzzy entropy and conditioning, *Information Sciences*, vol. 40, pp. 165–174, 1986.

21. A. De Luca and S. Termini, A definition of a nonprobabilistic entropy in the setting of fuzzy set theory, *Information and Control*, vol. 20, pp. 301–312, 1972.

22. S. K. Pal and P. K. Pramanik, Fuzzy measures in determining seed points in clustering, *Pattern Recognition Letters*, vol. 4, pp. 159–164, 1986.

23. N. R. Pal and S. K. Pal, Entropy: a new definition and its applications, *IEEE Transactions on Systems, Man, and Cybernetics,* vol. 21, pp. 1260–1270, 1991.

APPENDIX B

ARTIFICIAL NEURAL NETWORKS

Artificial neural networks (ANNs) are inspired by the biological nervous systems—the brain, which consists of a large number (approximately 10^{11}) of highly connected elements (approximately 10^4 connections per element) called *neurons*. The brain stores and processes the information by adjusting the linking patterns of the neurons. Artificial neural networks are signal processing systems that try to emulate the behavior of and ways of processing information in the biological nervous systems by providing a mathematical model of the combination of neurons connected in a network [1].

In an artificial neural network, artificial neurons are linked with each other through connections. Each connection is assigned a weight that controls the flow of information among the neurons. When the information is input into a neuron through the connections, it is summed up first and then undergoes a transformation by an activation function $f(x, w)$. The outputs of this activation function will be sent to other neurons or back to itself as input. A neuron that has three input connections and three output connections is shown in Figure B.1, where $x = (x_1, x_2, x_3)$ denotes the input vector and $w = (w_1, w_2, w_3)$ denotes the weight vector of the connections.

In artificial neural networks, input information is processed in parallel in the neurons. This improves the processing speed and the reliability of the neural network. Some advantages of ANN are summarized below:

- *Adaptive*. The network can modify its connection weights using some training algorithms or learning rules. By updating the weights, the ANN can optimize

Foundations of Soft Case-Based Reasoning.   By Sankar K. Pal and Simon C. K. Shiu
ISBN 0-471-08635-5   Copyright © 2004 John Wiley & Sons, Inc.

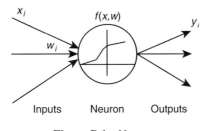

Figure B.1   Neuron.

its connections to adapt to the changes in the environments. This is the most important characteristic of ANN.

- *Parallel.* When the information is input into the ANN, it is distributed to different neurons for processing. Neurons can work in parallel and synergetically if they are activated by the inputs. In this arrangement, the computing power of the neural network is fully utilized and the processing time is reduced.

- *Rugged.* If one of the neurons fails, the weights of the connections can be adjusted for preserving the performance of the ANN. The working neurons will establish a stronger connection with each other, while the connections to the failed neuron will be weakened. By doing so, the reliability of the ANN is improved.

B.1   ARCHITECTURE OF ARTIFICIAL NEURAL NETWORKS

The architectures of ANNs can be classified into two categories based on the connections and topology of neurons:

- *Feedforward networks.* The inputs travel in only one direction, from input to output layer, and no feedback is allowed. Figure B.2 depicts a three-layered

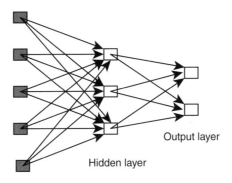

Figure B.2   Simple feedforward ANN.

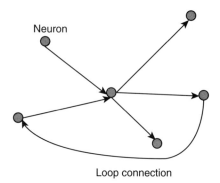

Figure B.3   Simple recurrent ANN.

feedforward ANN, where the information presented to the ANN travels from the input layer to the output layer following the direction of the arrows (i.e., from left to right). No loop or feedback occurs in the same layer. The recurrent ANNs are different from the feedforward ANNs because they consist of loops or feedbacks.

- *Recurrent (or feedback) networks.* The inputs can travel in both directions; loop is allowed. Figure B.3 depicts a simple recurrent ANN that has a loop in the connections. Before using an ANN, it needs to be trained. In the training phase, the weights are adjusted using some gradient-based algorithms or predefined learning rules. After training the ANN successfully, it can be used for problem solving.

B.2   TRAINING OF ARTIFICIAL NEURAL NETWORKS

There are principally three ways of training ANNs: supervised, unsupervised, and reinforcement training. In *supervised training*, weight modification is carried out by minimizing the difference between the ANN outputs and the expected outputs. In unsupervised training, weight modification is driven by the inputs. The weights are trained with some predefined learning rules that determine how to modify the weights. For example, the simple Hebb *learning rule* states that if two neighbor neurons show similar outputs, their connection will be strengthened (see Fig. B.4) as shown by

$$\Delta w_{ij} = \lambda y_i y_j \tag{B.1}$$

Figure B.4   Simple Hebb rule.

where Δw_{ij} is the modification of the weight and y_i and y_j are the output of the ith and jth node, respectively. λ is the step length, which is a small positive number, controlling the learning rate. In equation (B.1), the change in weight between the ith and jth nodes is proportional to the product of their outputs. If the outputs y_i and y_j are both positive or are both negative, the weight is increased and the link is strengthened. The system modifies its weights until the weight space is stabilized or comes to an oscillating state.

Reinforcement training is similar to supervised training, except that the training samples are obtained through the use of outputs from the neural networks. If the feedback of the output is successful, the input–output pair is stored as a training sample, and no modification is performed on the weight vector. Otherwise, the output will be repaired using some domain knowledge, which may be expressed in the form of a decision table or some IF–THEN rules. The repaired input–output pair will be stored and used as the training sample to modify, using supervised training, the weight vectors. This reinforcement training process is shown in Figure B.5.

In many applications, ANNs using supervised training, such as the back-propagation (BP) neural network and the radial basis function (RBF) neural network, are used to model the input–output relations of some complex systems, where an explicit mathematical model is hard to establish. ANNs using unsupervised training such as the Kohonen neural network, on the other hand, are suitable for data clustering.

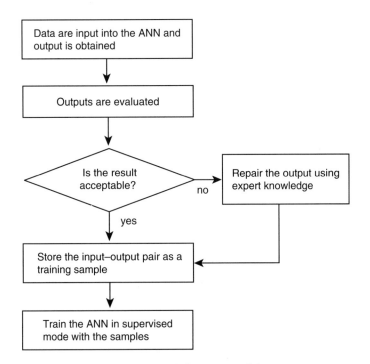

Figure B.5   Reinforcement training.

Neural networks using reinforcement training can be used when there are not enough training samples.

B.3 ANN MODELS

In this section we introduce four types of ANN models that are most commonly studied and used: the single-layered perceptron, multilayered perceptron, RBF neural network, and Kohonen neural network.

B.3.1 Single-Layered Perceptron

A single-layered perceptron (SLP) is the simplest neural network that has only one layer. In the one-output SLP, it consists of one neuron with many inputs. This type of neural network can be used for classification in the sample space that is linearly separable. As shown in Figure B.6, the perceptron has a set of weights, w_j, $j = 1, 2, \ldots, n$, and a threshold or bias Th. Given an input vector $x = [x_1, x_2, \ldots, x_n]^\mathrm{T}$, the net input to the neuron is

$$v = \sum_{j=1}^{n} w_j x_j - \mathrm{Th} \tag{B.2}$$

The output y of the perceptron is $+1$ if $v > 0$, or 0 otherwise. $+1$ and 0 represent two classes in a binary classification. The linear equation

$$\sum_{j=1}^{n} w_j x_j - \mathrm{Th} = 0 \tag{B.3}$$

represents the decision boundary that separates the space into two regions. Rosenblatt [2] developed a methodology for training the weights and threshold for classification. The following steps describe how to train the weights of the perceptron:

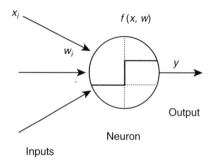

Figure B.6   Perceptron.

Step 1. Initialize the weights and threshold as small random numbers.

Step 2. Input the training vector $[x_1, x_2, \ldots, x_n]^T$ and evaluate the output of the neuron.

Step 3. Adjust the weights with equation (B.4) as

$$w_j(t+1) = w_j(t) + \lambda(\hat{y} - y)x_j \qquad (B.4)$$

where $\hat{y}$ is the desired output, t the iteration number, and λ $(0.0 < \lambda < 1.0)$ the step length.

Rosenblatt has proved that if the training patterns are drawn from a linearly separable sample space, the perceptron learning procedure converges after a finite number of iterations. However, if the sample space is not linearly separable, the perceptron fails. This leads to the development of the multilayer perceptron, discussed in the next section.

B.3.2   Multilayered Perceptron Using a Back-Propagation Algorithm

A multilayered perceptron (MLP) [3] is a network of more than one layer of neurons. The outputs of one layer become the inputs of the next layer and there is no connection among the nodes in the same layer. This type of neural network is capable of classifying the nonlinear separable data.

The training process is carried out using the gradient descent method. Initially, the training samples are preprocessed (e.g., normalized) and the weight space is initialized with random numbers. A sample is then selected from the training set, and its input is sent to the input layer and fed forward to the output layer through the hidden layer(s). The output of the ANN is compared with the desired output, and an error function that takes the weights as variables is established to measure the difference. To adjust the weights, the error function is minimized using the gradient descent method. In calculating the gradient, the derivative of the error function with respect to the weight vector of a certain node can be worked out using a recursive equation that back-propagates the weight adjustment from the output layer to the input layer. This process will repeat for many cycles until the value of the error function becomes smaller than a user-specified threshold. This entire process may be repeated using other samples for further training of the network. After getting trained with all the samples, the neural network can be seen as a model of the input–output relation of the sample space. It can then be used for problem solving. This is explained below with respect to a multilayered perceptron (Fig. B.7).

Here j, $j+1$, and L denote the jth layer, the $(j+1)$th layer, and the Lth layer (output layer), respectively; i and m denote the ith node in the jth layer and the mth node in the $(j+1)$th layer, respectively; $N(j)$ is the number of nodes in the jth layer; $y_i^L(k)$ is the output of the ith node in the Lth layer in the kth training cycle; $\hat{y}_i$ is the expected output of the ith node in the ouput layer; s_i^j and w_i^j are the

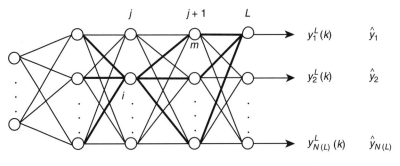

Figure B.7   General structure of a multilayer perceptron.

weighted sum and the weight vector of the ith node in the jth layer; and w_{mi}^{j+1} is the weight between the mth node in the $(j+1)$th layer and the ith node in the jth layer.

The sum of the squares of the errors in the output layer,

$$\mathrm{Er}(k) = \sum_{v=1}^{N(L)} [y_v^L(k) - \hat{y}_v]^2 \qquad (\mathrm{B}.5)$$

is used as the error function. The weights of the network are adjusted by minimizing this function. To minimize $\mathrm{Er}(k)$ the gradient descent method is used. The gradient of $\mathrm{Er}(k)$ with respect to the weight vector of the ith node in the jth layer is

$$\frac{\partial \mathrm{Er}(k)}{\partial w_i^j(k)} = 2 \sum_{v=1}^{N(L)} [y_v^L(k) - \hat{y}_v] \frac{\partial y_v^L(k)}{\partial w_i^j(k)} \qquad (\mathrm{B}.6)$$

To work out equation (B.6), we use the chain rule:

$$\frac{\partial y_v^L(k)}{\partial w_i^j(k)} = \frac{\partial y_v^L(k)}{\partial s_i^j(k)} \frac{\partial s_i^j(k)}{\partial w_i^j(k)} = \frac{\partial y_v^L(k)}{\partial s_i^j(k)} O^{j-1} \qquad (\mathrm{B}.7)$$

As O^{j-1} is the output vector from the $(j-1)$th layer, which is already known, we only need to solve equation (B.7). Since $\partial y_l^j(k)/\partial s_i^j(k) = 0$, when $l \neq i$, the formula is simplified to

$$\frac{\partial y_v^L(k)}{\partial s_i^j(k)} = \sum_{m=1}^{N(j+1)} \frac{\partial y_v^L(k)}{\partial s_m^{j+1}(k)} \frac{\partial s_m^{j+1}(k)}{\partial s_i^j(k)} = \sum_{m=1}^{N(j+1)} \frac{\partial y_v^L(k)}{\partial s_m^{j+1}(k)} \left[\sum_{l=1}^{N(j)} \frac{\partial w_{ml}^{j+1}(k) y_l^j(k)}{\partial s_i^j(k)} \right]$$

$$= \frac{\partial y_i^j(k)}{\partial s_i^j(k)} \sum_{m=1}^{N(j+1)} \frac{\partial y_v^L(k)}{\partial s_m^{j+1}(k)} w_{mi}^{j+1}(k) \qquad (\mathrm{B}.8)$$

Denote $\Delta^j_{iv}(k) = (y^L_v(k) - \hat{y}_v)\partial y^L_v(k)/\partial s^j_i(k)$; thus from equation (B.8) we have

$$\Delta^j_{iv}(k) = \frac{\partial y^j_i(k)}{\partial s^j_i(k)} \sum_{m=1}^{N(j+1)} \Delta^{j+1}_{mv}(k) w^{j+1}_{mi}(k) \tag{B.9}$$

Equation (B.9) is called the *delta rule*. Since the delta value in the Lth layer is

$$\Delta^L_{iv}(k) = \begin{cases} (y^L_v(k) - \hat{y}_v)\dfrac{\partial y^L_v(k)}{\partial s^L_v(k)} & i = v \\ 0 & i \neq v \end{cases} \tag{B.10}$$

all the delta values can be calculated recursively with equations (B.9) and (B.10). Therefore,

$$\frac{\partial \mathrm{Er}(k)}{\partial w^j_i(k)} = 2\sum_{v=1}^{N(L)} \Delta^j_{iv}(k) O^{j-1}(k) \tag{B.11}$$

So, to adjust the weight space iteratively,

$$w^j_i(k+1) = w^j_i(k) - \lambda \sum_{v=1}^{N(L)} \Delta^j_{iv}(k) O^{j-1}(k) \tag{B.12}$$

where λ is the step length, which is a small positive number, predefined by the user, to control the learning rate.

B.3.3 Radial Basis Function Network

The RBF neural network is a type of feedforward ANN with an architecture similar to that of a three-layered MLP. Figure B.8 depicts a RBF neural network with inputs of n-dimension and outputs of m-dimension. The differences between the RBF

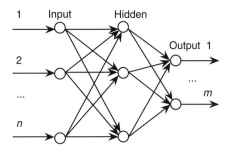

Figure B.8 Radial basis function network.

neural network and the MLP can be summarized as: (1) in RBF NN, the input is sent to the hidden layer without being weighted; (2) the hidden layer uses a radial basis function (e.g., the Gaussian function); and (3) the final layer outputs the weighted sum of the hidden layer's outputs without transformation.

As shown in Figure B.8, $x = (x_1, x_2, \ldots, x_n) \in R^n$ and $y = (y_1, y_2, \ldots, y_m) \in R^m$ are the input and output respectively, and h is the number of the nodes in the hidden layer. Thus, the output v_j of the jth node in the hidden layer is

$$v_j = \exp\left[-\frac{(x - c_j)^{\mathrm{T}}(x - c_j)}{\sigma_j^2}\right], \qquad j = 1, 2, \ldots, h \qquad (\text{B.13})$$

Here $c_j \in R^n$ is the center of the Gaussian function for node j and σ_j is the width of the jth center. The output y_j of the jth node in the output layer is

$$y_j = w_j^{\mathrm{T}} v, \qquad j = 1, 2, \ldots, m \qquad (\text{B.14})$$

where w_j is the weight vector of the jth node in the output layer and $v = (v_1, v_2, \ldots, v_h) \in R^h$ is the output vector from the hidden layer. To train the neural network, the parameters c_j, σ_j, and w_j need to be identified. By establishing an error function Er and using the gradient descent method, these parameters can be determined:

$$Er = \frac{1}{2}\sum_{j=1}^{m}(y_j - \hat{y}_j)^2 \qquad (\text{B.15})$$

where y_j and $\hat{y}_j$ are the neural network output and the desired output of the jth node in the final layer, respectively. After initializing the parameters, the parameters of the neural network can be updated using

$$w_j(k+1) = w_j(k) - \lambda_1 \frac{\partial Er(k)}{\partial w_j(k)}, \qquad j = 1, 2, \ldots, m \qquad (\text{B.16})$$

$$c_j(k+1) = c_j(k) - \lambda_2 \frac{\partial Er(k)}{\partial c_j(k)}, \qquad j = 1, 2, \ldots, h \qquad (\text{B.17})$$

$$\sigma_j(k+1) = \sigma_j(k) - \lambda_3 \frac{\partial Er(k)}{\partial \sigma_j(k)}, \qquad j = 1, 2, \ldots, h \qquad (\text{B.18})$$

Here k denotes the kth cycle in the training process; h and m are the number of nodes in the hidden layer and the output layer, respectively; and λ_1, λ_2, and λ_3 are the step lengths, controlling the learning rates, corresponding to the parameters weight, center, and width.

B.3.4 Kohonen Neural Network

The Kohonen neural network [4,5] is a type of competitive learning network. Unlike other neural networks, such as BP and RBF, which require the input–output sample pairs for the gradient-based training, the Kohonen network is trained based on competitive learning rules using only the input samples. Kohonen network can be used in unsupervised data clustering. Figure B.9 depicts the structure of a Kohonen network that generates four clusters from l three-dimensional input points. In Figure B.9, all the input units i, $i = 1, 2, 3$, are connected to all the output competitive neurons j, $j = 1, 2, 3, 4$, with the weight w_{ij}. The number of the input neurons is equal to the dimension of the input patterns, while the number of the output neurons corresponds to the number of the clusters.

The center of a cluster is represented by the weight vector connected to the corresponding output unit. To determine the centers, the weight vectors are trained using inputs that are sequentially presented. The following steps describe how to find cluster centers by training the neural network:

Step 1. Present a three-dimensional input vector x to the neural network and select the winning weight vector w_c that satisfies

$$||x - w_c|| = \min_i ||x - w_i||, \qquad i = 1, 2, 3, 4 \qquad \text{(B.19)}$$

where $|| \cdot ||$ is defined as the Euclidean distance.
Step 2. Update the winning vector and all of its neighbors with

$$\Delta w_i = \lambda(x - w_i), \qquad i \in N_c \qquad \text{(B.20)}$$

where N_c denotes a set of nodes lying in a window centred at w_c, and λ is the step length, which is a small positive number used to controll the learning rate. To achieve a better convergence, λ should decrease gradually with each iteration of weight adjustment.

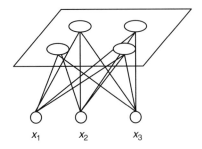

Figure B.9   Kohonen network.

The width or radius of N_c can be time variable; in fact, for good global ordering, it has turned out experimentally to be advantageous to let N_c be very wide in the beginning and shrink monotonically with time. This is because a wide initial N_c, corresponding to a coarse spatial resolution in the learning process, first induces a rough global order in the w_i values, after which narrowing of N_c improves the spatial resolution of the map; the acquired global order, however, is not destroyed later. This allows the topological order of the map to be formed.

By repeating the above-mentioned two learning steps for several cycles, the weight vectors will converge to the cluster centers of the data set. Therefore, the trained Kohonen network provides twofold quantization: (1) the l input points are quantized to four prototypes, preserving the data (density) distribution; and (2) the three-dimensional input space is reduced to a two-dimensional map reflecting the topological distribution of the data.

REFERENCES

1. S. K. Pal and S. Mitra, *Neuro-Fuzzy Pattern Recognition: Methods in Soft Computing*, Wiley, New York, 1999.

2. F. Rosenblatt, The perceptron: a probabilistic model for information storage and organization in the brain, *Psychological Review*, vol. 65, pp. 386–408, 1958.

3. D. E. Rumelhart and J. L. McClelland (eds.), *Parallel Distributed Processing: Explorations in the Microstructures of Cognition*, vol. 1, MIT Press, Cambridge, MA, 1986.

4. T. Kohonen, Analysis of a simple self-organizing process, *Biological Cybernetics*, vol. 44, pp. 135–140, 1982.

5. T. Kohonen, *Self-Organization and Associative Memory*, Springer-Verlag, Berlin, 1989.

APPENDIX C

GENETIC ALGORITHMS

The genetic algorithm (GA), conceived by John Holland in the early 1970s, is an adaptive and heuristic-based search algorithm inspired by natural evolution. It is based on biological evolutionary mechanisms such as natural selection, crossover, and mutation. GAs are becoming increasingly popular because of their wide uses in many applications, especially in optimization. In the following, the basic concepts of GAs are described; readers can refer to [1–11] for more detailed information.

C.1 BASIC PRINCIPLES

Genetic algorithms [4,7] are adaptive and robust computational procedures modeled on the mechanics of natural genetic evolutionary systems. They are viewed as randomized, yet structured, search and optimization techniques. GAs are executed iteratively on a set of coded solutions, called a *population*, with three basic operators: selection, crossover, and mutation. To determine the optimal solution of a problem, a GA starts from a set of assumed solutions (i.e., chromosomes) and evolves better sets of solutions over a sequence of iterations. In each generation (or iteration) the objective function (i.e., fitness) determines the suitability of each solution, and based on these values, some of the solutions (which are called *parent chromosomes*) are selected for reproduction. The number of offspring reproduced by an individual parent is proportional to its fitness value; thus the good (highly fitted) chromosomes are selected and the bad ones are eliminated.

Foundations of Soft Case-Based Reasoning. By Sankar K. Pal and Simon C. K. Shiu
ISBN 0-471-08635-5 Copyright © 2004 John Wiley & Sons, Inc.

The following explains how GAs mimic some of the principles observed in natural evolution [2]:

- Evolution takes place at the level of chromosomes rather than on the living beings themselves. Therefore, GAs operate on the encoded possible solutions (called *chromosomes*) of the problems.
- Nature obeys the principle of Darwinian "survival of the fittest." The chromosomes with high fitness values will, on average, reproduce more often than those with low fitness values. In GAs, selection of chromosomes also tries to mimic this principle.
- Biological entities adapt themselves in response to changes in the natural environment. Similarly, GAs use the objective function to evaluate chromosomes, and those having high fitness values are selected in the selection process.

C.2 STANDARD GENETIC ALGORITHM

The standard genetic algorithm is the one most used in applications of GAs. It consists of three primary steps: selection, crossover, and mutation. Given an initial population of solutions (i.e., choromosomes), the algorithm is carried out iteratively to find an optimal solution(s) that satisfies user requirements. The pseudocode of this algorithm is shown in Figure C.1.

The basic concepts and operations in the standard GA, include the following:

1. *Chromosome.* The solution of a problem is encoded as chromosomes before being used in genetic operations. For example, if the integer 240 is a solution of a problem, it can be encoded as a binary string 11110000. Each bit in the string is called a *gene* and the entire string is called a *chromosome* or an *individual.*

```
Input: An initial population of n coded solutions (chromosomes).
Output: The solution with the highest fitness value.
Begin
          i = 0;
      While (no satisfying solution is found in the ith generation)
            Fitness_computation;
            Selection;
            Crossover;
            Mutation;
            i = i + 1;
      EndWhile
      Return The solution with the highest fitness value in the ith
      generation;

End
```

Figure C.1   Standard algorithm.

2. *Population.* A group of chromosomes/individuals constitute a population. Since each chromosome represents a solution, a population is thus a solution set of the problem.

3. *Fitness value.* Each individual (e.g., x_i), corresponds to a fitness value $f(x_i)$ (i.e., the fitness of x_i). The larger the value of $f(x_i)$, the better the fitness of an individual to the environment. Individuals with large fitness values will have more chances of being selected for reproduction than those with small fitness values. In real-life applications, the fitness function is constructed depending on the problem domain.

4. *Selection operation.* The objective of selection is to choose the good individuals from the population and use them for reproduction. To mimic the "survival of the fittest" principle, individuals with larger fitness values will have more chances of being selected for reproduction. To determine the chance/possibility for an individual, the following equation can be used:

$$\text{Selecprob} = \frac{f(x_i)}{\sum_{i=1}^{n} f(x_i)} \tag{C.1}$$

where $f(x_i)$ is the fitness value of the individual x_i and Selecprob is the selection probability for x_i.

5. *Crossover operation.* Living beings reproduce new generations by performing the crossover operations on their chromosomes. This concept is also introduced into GA. To make crossover between two chromosomes a and b selected from the population, a cutting point (cross site) is selected. The genes behind this cutting point will then be exchanged to generate new chromosomes a' and b':

$$\downarrow \qquad\qquad\qquad\qquad\qquad\qquad \downarrow$$

$$a = 11000\ 10101\ 01000\ 01111\ 10001 \quad \rightarrow \quad b = 10001\ 01110\ 11101\ 00110\ 10100$$
$$a' = 11000\ 10101\ 01101\ 00110\ 10100 \quad \rightarrow \quad b' = 10001\ 01110\ 11000\ 01111\ 10001$$

where $\downarrow$ marks the position of the cutting point. The last 10 bits of a and b are exchanged with each other.

6. *Mutation operation.* Besides the crossover operation, the mutation operation is also important for reproducing individuals with new features. This operation is performed on some randomly selected genes of a chromosome. For example, a mutation may occur on b' by inverting its last bit from 1 to 0, thereby generating a new chromosome b'':

$$b' = 10001\ 01110\ 11000\ 01111\ 1000\mathbf{1} \rightarrow b'' = 10001\ 01110\ 11000\ 01111\ 1000\mathbf{0}$$

So far, GAs have been used successfully, especially in optimization (e.g., the traveling salesman problem, the scheduling problem, and the nonlinear programming problem). In the next section, two examples are given to illustrate the use of GAs.

C.3   EXAMPLES

The first example shows how to find the maximum point of a real value function. The second one describes how to use the GA to solve the traveling salesman problem (TSP).

C.3.1   Function Maximization

Given a two-dimensional function [12],

$$f(x, y) = (1 - x)^2 e^{-x^2 - (y+1)^2} - (x - x^3 - y^3)e^{-x^2 - y^2}, \qquad (x, y) \in [-3, 3] \times [-3, 3] \tag{C.2}$$

find its maximum point.

The following steps show how to solve this problem using GAs.

Step 1.   Initialize the population of the chromosomes. Each chromosome is encoded as a binary string with x-part and y-part. A sample chromosome c_1 takes the form shown in Figure C.2. Break the chromosome into an x-part and a y-part and convert them to the decimal numbers. For the sample chromosome c_1, see Figure C.3. Since the value of the 8-bit decimal ranges from 0 to 255, which may exceed the boundary of $[-3,3]$, a proportion conversion is made. The proportion factor is

$$\frac{3 - (-3)}{255} = 0.0235294$$

To obtain the values of x and y, we multiply the decimal value with the proportion factor 0.0235294 and then subtract 3 from the results; thus

$$x_1 = (138)_{10} \times 0.0235294 - 3 = 0.2470588$$
$$y_1 = (59)_{10} \times 0.0235294 - 3 = -1.6117647$$

| x | 1 | 0 | 0 | 0 | 1 | 0 | 1 | 0 | | 0 | 0 | 1 | 1 | 1 | 0 | 1 | 1 | y |

Figure C.2   Chromosome.

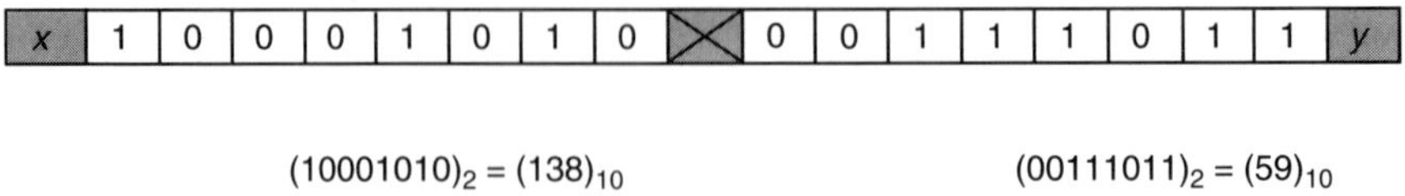

| x | 1 | 0 | 0 | 0 | 1 | 0 | 1 | 0 | ✕ | 0 | 0 | 1 | 1 | 1 | 0 | 1 | 1 | y |

$(10001010)_2 = (138)_{10}$ $(00111011)_2 = (59)_{10}$

Figure C.3   Chromosome is broken into x-part and y-part.

Step 2. Input x and y to calculate the fitness of the chromosomes, and then compute the selection probability of each chromosome. For the sample chromosome we have

$$f(c_1) = f(x_1, y_1) = f(0.2470588, \; -1.6117647) = 0.05737477$$

Other fitness values can be calculated similarly. The selection probability of the chromosome is

$$\text{Selecprob} = \frac{f(c_1)}{\sum_{i=1}^{n} f(c_i)} \tag{C.3}$$

where n is the number of the chromosomes of the population and c_i is the ith chromosome. The number of times that c_i is being selected for reproduction is

$$\text{Selectimes} = n \cdot \text{Selecprob} \tag{C.4}$$

Step 3. Randomly select a pair of chromosomes for performing crossover based on the crossover probability, and set the cross site (cutting point) randomly. Perform crossover operation by exchanging the genes that lies behind the cutting point.

Step 4. Set the mutation probability (Mutaprob) to a value, say, 0.001. The number of the genes that will be changed in the population is then

$$\text{Mutagenesu } m = 16n \times 0.001$$

Convert the bits from 1 to 0, or 0 to 1, to create a new generation of chromosomes.

Step 5. Repeat steps 1 to 4 until the fitness values of the chromosomes satisfy the user requirement. The individual with the highest fitness value thus represents the maximum point of the real value function.

The different states of the population during the process of searching the maximum point are demonstrated in Figures C.4 to C.7. Figure C.4 shows the initial

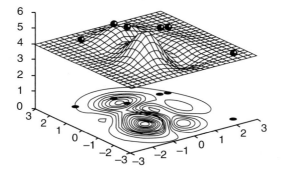

Figure C.4 Initial generation. (From [12].)

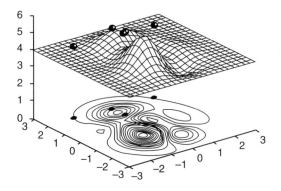

Figure C.5 First generation. (From [12].)

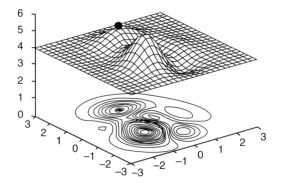

Figure C.6 Local maximum. (From [12].)

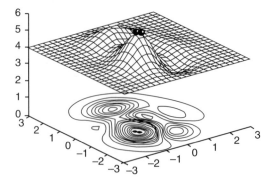

Figure C.7 Global maximum. (From [12].)

state of the population, where the individuals are far from the maximum point. After many cycles of GA operations, as shown in Figure C.5, the individuals get closer to the maximum point. Figures C.6 and C.7 show two final states that are possible for the individuals. In Figure C.6 the individuals reach to a local maximum point; in Figure C.7 the individuals arrive at the point of the global maximum. Whether or not individuals can converge to the global maximum point depends on many factors, including the initial value assigned to the chromosomes. Readers may refer to [2] for detailed information on the convergence properties of GAs.

C.3.2 Traveling Salesman Problem

The traveling salesman problem is stated as follows: "Given the cost of travel between n cities, how should the salesman plan his itinerary for the minimum total cost of the entire tour?" [7]. Each tour is a permutation of the cities visited in order. For a given permutation of the cities $pm_k = (c_{1k}, c_{2k}, \ldots, c_{nk})$, the total cost function is defined as

$$\text{Totalcost}(pm_k) = \sum_{i=1}^{n-1} d(c_{ik}, c_{i+1k}) + d(c_{nk}, c_{1k}) \qquad (\text{C.5})$$

where d is the cost function between two cities. The fitness function of the TSP is

$$f(pm_k) = \frac{1}{\text{Totalcost}(pm_k)} \qquad (\text{C.6})$$

It can be seen that a permutation of the n cities is a candidate solution, so there are a total of $n!$ patterns of permutations. These patterns form a very large data set. Searching for the qualified solutions in such a huge data set is time consuming or even impossible. In this situation, GAs can be applied to get the optimized solution more efficiently. The following steps describe how to use GAs to solve the TSP problem:

Step 1. Encode the traveling routes as the chromosomes. Each chromosome is represented as a string that shows the sequence of the visited cities. For example, if there are nine cities and the traveling order is

$$5 \to 1 \to 7 \to 8 \to 9 \to 4 \to 6 \to 2 \to 3$$

the solution is encoded as (5 1 7 8 9 4 6 2 3). A group of chromosomes in this form will be initialized to create a population.

Step 2. Use the fitness function [equation (C.6)] to calculate the fitness of the chromosomes. Then make selections using

$$\text{Selecprob}(pm_k) = \frac{f(pm_k)}{\sum_{i=1}^{n} f(pm_i)} \qquad (\text{C.7})$$

Step 3. Set cutting points randomly and perform crossover operations. For example, for a given couple pm_1 and pm_2, the crossover is performed as follows:

$$pm_1 = (1\ 2\ 3\ \mathbf{4\ 5\ 6\ 7}\ 8\ 9) \rightarrow (1\ 2\ 3\ \mathbf{1\ 8\ 7\ 6}\ 8\ 9)$$
$$pm_2 = (4\ 5\ 2\ \mathbf{1\ 8\ 7\ 6}\ 9\ 3) \rightarrow (4\ 5\ 2\ \mathbf{4\ 5\ 6\ 7}\ 9\ 3)$$

Gene mapping: $4 \rightarrow 1$; $5 \rightarrow 8$; $6 \rightarrow 7$; $7 \rightarrow 6$

$$\underset{4}{(1}\ 2\ 3\ \underline{\mathbf{1}}\ 8\ 7\ 6\ \underset{5}{8}\ 9) \rightarrow (4\ 2\ 3\ 1\ 8\ 7\ 6\ 5\ 9)$$
$$\underset{1\ 8}{(4}\ 5\ 2\ \mathbf{4}\ \underline{\mathbf{5}}\ 6\ 7\ 9\ 3) \rightarrow (1\ 8\ 2\ 4\ 5\ 6\ 7\ 9\ 3)$$

where the $\downarrow$ marks the cutting point. First we exchange genes between the cutting points. This derives a mapping between genes in the exchanged section. Since a valid permutation should not contain the duplicated genes (i.e., no loop is allowed in the traveling route), we replace the duplicated genes with their corresponding mappings. For example, the chromosome

$$1\ 2\ 3\ 1\ 8\ 7\ 6\ 8\ 9$$

has duplicated genes 1 and 8. These two genes are replaced by their mappings 4 and 5, respectively.

Step 4. Mutate the chromosomes to create a new generation. For example, we can reverse the order of the genes between the cutting points for mutation.

$$(4\ 2\ 3\ 1\ 8\ 7\ 6\ 5\ 9) \rightarrow (4\ 2\ 3\ 6\ 7\ 8\ 1\ 5\ 9)$$
$$(1\ 8\ 2\ 4\ 5\ 6\ 7\ 9\ 3) \rightarrow (1\ 8\ 2\ 7\ 6\ 5\ 4\ 9\ 3)$$

Other couples are processed in the same way. The final solution should be an optimized or a nearly optimized route that provides an itinerary with low traveling cost.

It should be noted that there are many other encoding methods, crossover operators, and mutation operators that can be used for the TSP problem. Interested readers can refer to [34] for more details.

REFERENCES

1. D. Bhandari, N. R. Pal, and S. K. Pal, Directed mutation in genetic algorithms, *Information Sciences*, vol. 79, p. 251, 1994.

2. L. Davis (ed.), *Handbook of Genetic Algorithms*, Van Nostrand Reinhold, New York, 1991.

3. L. J. Eshelman, R. A. Caruana, and J. Schaffer, Biases in the crossover landscape, in *Proceedings of the Third International Conference on Genetic Algorithms*, George Mason University, Fairfax, VA, Morgan Kaufmann, San Francisco, pp. 10–19, 1989.

4. D. E. Goldberg, *Genetic Algorithms in Search, Optimization and Machine Learning*, Addison-Wesley, Reading, MA, 1989.

5. J. H. Holland, *Adaptation in Natural and Artificial Systems*, University of Michigan Press, Ann Arbor, MI, 1975.

6. V. Maniezzo, Genetic evolution of the topology and weight distribution of neural networks, *IEEE Transactions on Neural Networks*, vol. 5, no. 1, pp. 39–53, 1994.

7. Z. Michalewizq, *Genetic Algorithms + Data Structure = Evolution Programs*, Springer-Verlag, Berlin, 1992.

8. G. Syswerda, Uniform crossover in genetic algorithms, in *Proceedings of the Third International Conference on Genetic Algorithms*, Morgan Kaufmann, San Francisco, pp. 2–9, 1989.

9. D. Whitley and J. Kauth, Genitor: a different genetic algorithm, in *Proceedings of the Rocky Mountain Conference on Artificial Intelligence*, Denver, CO, pp. 118–130, 1988.

10. D. Whitely, T. Starkweather, and C. Bogart, Genetic algorithms and neural networks: optimizing connections and connectivity, *Parallel Computing*, vol. 14, pp. 347–361, 1990.

11. S. K. Pal and P. P. Wang, *Genetic Algorithms for Pattern Recognition*, CRC Press, Boca Raton, FL, 1996.

12. M. Negnevitsky, *Artificial Intelligence: A Guide to Intelligent Systems*, Addison-Wesley, Reading, MA, 2002.

APPENDIX D

ROUGH SETS

Rough set theory, developed by Zdzislaw Pawlak in the early 1980s, provides a mathematical approach to deal with problems with uncertainty and vagueness [1–3]. It is often used to uncover irregularities and relationships in data. The main advantages of this theory [4] are that it:

- Provides efficient algorithms for discovering hidden patterns in data
- Identifies relationships that would not be found while using statistical methods
- Allows the use of both qualitative and quantitative data
- Finds the minimal sets of data that can be used for classifications (i.e., data reduction)
- Evaluates the significance of data
- Generates sets of decision rules from data

Before we provide the definition of rough set, we first present some of the basic concepts and mathematical definitions related to rough set theory. For more details, readers can refer to [1–20].

D.1 INFORMATION SYSTEMS

In applying rough sets, the data used are usually represented in a flat table as follows: Columns represent the attributes, rows represent the objects, and every

Foundations of Soft Case-Based Reasoning. By Sankar K. Pal and Simon C. K. Shiu
ISBN 0-471-08635-5 Copyright © 2004 John Wiley & Sons, Inc.

TABLE D.1   Example of an Information System

Object (e.g., Department Store)	Attribute 1 (e.g., E)	Attribute 2 (e.g., Q)	Attribute 3 (e.g., L)
1	High	Good	No
2	Medium	Good	No
3	Medium	Good	No
4	Low	Average	No
5	Medium	Average	Yes
6	High	Average	Yes

cell contains the attribute value for the corresponding objects and attributes. In rough set terminology, such tables are called *information systems*. More formally, an information system is a pair $\hat{S} = (U, A)$, where U is a nonempty finite set of objects called a *universe* and A is a nonempty finite set of attributes. Every attribute Attr $\in A$ has an attribute value set V_{Attr} associated with it (i.e., Attr: $U \rightarrow V_{\text{Attr}}$). An example of an information system is shown in Table D.1. In this information system, there are six objects (i.e., some department stores), which are characterized by three attributes: E, Q, and L, where E represents the experience of the sales personnel, Q the perceived quality of the merchandises, and L refers to the availability of a nearby train station. From Table D.1, department stores 2 and 3 have exactly the same attribute values, and therefore the two stores are indiscernible using the attributes available.

In many real-life classification problems, the outcome (or class label) is already known. This class label is referred as a *decision attribute*. An information system that includes the decision attribute is called a *decision system*, which is denoted as Ts $= (U, A \cup \{d^*\})$, where $d^* \notin A$ is the decision attribute. The elements of A are called *conditional attributes*. More generally, a decision system is represented by Ts $= (U, \text{Tc}, \text{Td})$, where Tc is the set of condition attributes and Td is the set of decision attributes. Table D.2 shows an example of a decision system. This table is the same as Table D.1 except that it has the decision attribute Pf, which represents the profit status of the department stores. In this table, department stores 2 and 3 have the same condition attribute values but with different decision attribute values (i.e., loss and profit, respectively).

TABLE D.2   Example of a Decision System

Store	E	Q	L	Pf
1	High	Good	No	Profit
2	Medium	Good	No	Loss
3	Medium	Good	No	Profit
4	Low	Average	No	Loss
5	Medium	Average	Yes	Loss
6	High	Average	Yes	Profit

D.2 INDISCERNIBILITY RELATION

Given an information system, each subset of attributes can be used to partition the objects into clusters. The objects in the same cluster have the same attribute values. Therefore, these objects are indiscernible (or similar) based on the available information. An indiscernibility relation can be defined to describe this property. Before doing so, some relevant concepts are given first.

Definition D.1. An *ordered pair* is a set of a pair of objects with an order associated with them. If objects are represented by x and y, we write an ordered pair as (x,y) or (y,x). In general (x,y) is different from (y,x).

Definition D.2. A binary *relation* from a set X to a set Y is a set of ordered pairs (x,y) where x is an element of X and y is an element of Y.

When an ordered pair (x,y) is in a relation R, we write $x\,R\,y$ or $(x,y) \in R$. It means that element x is related to element y in relation R. When $X = Y$, a relation from X to Y is called a (binary) relation on Y.

Definition D.3. The set of all ordered pairs (x,y) is called the *Cartesian product* of X and Y, denoted as $X \times Y$, where x is an element of X and y is an element of Y.

Thus for a binary relation R from X to Y, it is a subset of Cartesian product $X \times Y$ denoted as $R \subseteq X \times Y$.

Definition D.4. A binary relation $R \subseteq X \times X$, which is reflexive (i.e., an object is in relation with itself, denoted as $x\,R\,x$), symmetric (if $x\,R\,y$, then $y\,R\,x$) and transitive (if $x\,R\,y$ and $y\,R\,z$, then $x\,R\,z$) is called an *equivalence relation*. The equivalence class of an element $x \in X$, denoted as $[x]_R$, consists of all objects $y \in X$ such that $x\,R\,y$. The family of all equivalence classes of R is denoted by U/R.

Now let us provide a definition of the indiscernibility relation.

Definition D.5. Given an information system $\hat{S} = (U, A)$, with any $B \subseteq A$, there is an associated equivalence relation I_B:

$$I_B = \{(x, y) \in U \times U | \forall\, \text{Attr} \in B, \text{Attr}(x) = \text{Attr}(y)\} \qquad (D.1)$$

I_B is called the *B-indiscernibility relation*.

If $(x, y') \in I_B$, the objects x and y' are indiscernible from each other by attributes from B. The family of all equivalence classes of I_B (i.e., the partition determined by B) will be denoted by U/I_B, or simply U/B; an equivalence class of I_B containing x will be denoted by $[x]_B$. Now we use the information system given in Table D.2 to illustrate the indiscernibility relation and the corresponding equivalence classes.

Here are six nonempty conditional attributes sets: $\{E\}$, $\{Q\}$, $\{L\}$, $\{E,Q\}$, $\{E,L\}$, $\{Q,L\}$, and $\{E,Q,L\}$. Considering the attribute set $\{E,Q\}$, by equation (D.1) we can get the $\{E,Q\}$-indiscernibility relation, $I_{\{E,Q\}}$, and the partition determined by it is as below:

$$U/I_{\{E,Q\}} = \{\{1\},\{2,3\},\{4\},\{5\},\{6\}\}$$

Considering $\{2,3\} \in I_{\{E,Q\}}$, objects 2 and 3 are indiscernible from each other by E and Q and belong to the same equivalence class.

Similarly for the other attribute sets, we can obtain their corresponding indiscernibility relations and equivalent classes. Below are partitions determined by the indiscernibility relations $I_{\{Q,L\}}$, $I_{\{E\}}$, and $I_{\{E,Q,L\}}$, respectively.

$$U/I_{\{Q,L\}} = \{\{1,2,3\},\{4\},\{5,6\}\}$$
$$U/I_{\{E\}} = \{\{1,6\},\{2,3,5\},\{4\}\}$$
$$U/I_{\{E,Q,L\}} = \{\{1\},\{2,3\},\{4\},\{5\},\{6\}\}$$

D.3 SET APPROXIMATIONS

Before describing the concept of rough sets, let us define set approximation.

Definition D.6. Consider an information system $\hat{S} = (U,A)$. With each subset $X \subseteq U$ and $B \subseteq A$, we associate two crisp sets

$$\underline{B}X = \{x|[x]_B \subseteq X\} \tag{D.2}$$
$$\overline{B}X = \{x|[x]_B \cap X \neq \emptyset\} \tag{D.3}$$

called the *B-lower* and *B-upper approximations* of X, respectively.

The set $\underline{B}X$ (or $\overline{B}X$) consists of objects, which surely (or possibly) belong to X with respect to the knowledge provided by B. The set $BN_B(X) = \overline{B}X - \underline{B}X$, called the *B-boundary region* of X, consists of those objects that cannot surely belong to X.

A set is said to be *rough* if the boundary region is nonempty with respect to B. That is, if $BN_B(X) \neq \emptyset$, the set X is called *rough* (i.e., inexact) with respect to B; and if $BN_B(X) = \emptyset$, the set X is *crisp* (i.e., exact) with respect to B, in contrast. The set $U - \overline{B}X$ is called the *B-outside region* of X, and it consists of objects that certainly cannot belong to X (on the basis of knowledge in B).

An example of set approximation is shown below. Let $X = \{x|\text{Pf}(x) = \text{Profit}\}$. From Table D.2 we then obtain $X = \{1,3,6\}$. Let us assume that $B = \{E,Q,L\}$; then

$$[1]_B = \{1\}, \quad [2]_B = [3]_B = \{2,3\}, \quad [4]_B = \{4\}, \quad [5]_B = \{5\}, \quad [6]_B = \{6\}$$

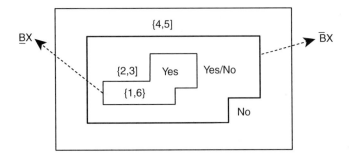

Figure D.1 Approximating the set of profit stores using three conditional attributes.

By the definitions of the lower and upper approximations, we have

$$\underline{B}X = \{1,6\}, \quad \overline{B}X = \{1,2,3,6\}, \quad BN_BX = \{2,3\}$$

It can be seen that the boundary region is not empty, and this shows that the set X is rough. Objects in the store set $\{1,6\}$ surely belong to those making profit, and the store set $\{1,2,3,6\}$ includes objects possibly making profit, where objects 2 and 3 included in the boundary region cannot be classified either as making a profit or having a loss. $U - \overline{B}X = \{4,5\}$ shows that stores 4 and 5 certainly do not belong to those that are making profit. Here the approximations of X are shown in Figure D.1.

A rough set can also be described numerically using the concept of *accuracy of approximation* as follows:

$$\alpha_B(X) = \frac{|\underline{B}X|}{|\overline{B}X|} \tag{D.4}$$

where $|X|$ denotes the cardinality of $X \neq \emptyset$, and $0 \le \alpha_B(X) \le 1$. If $\alpha_B(X) = 1$, X is crisp with respect to B. Otherwise [i.e., if $\alpha_B(X) < 1$], X is rough with respect to B.

D.4 ROUGH MEMBERSHIP

The degree of object x belonging to X can be described by rough membership function. This membership function quantifies the degree of relative overlap between the set X and the equivalence class $[x]_B$, and it is defined by

$$\mu_X^B : U \to [0, 1] \quad \text{and} \quad \mu_X^B(x) = \frac{|[x]_B \cap X|}{|[x]_B|} \tag{D.5}$$

If the attribute set B is given, for a certain subset $X \subseteq U$, the membership function can be determined, and correspondingly, the degree of each element $x \in U$ belonging to X is obtained.

Using a rough membership function, we can define the lower-approximation, upper-approximation, and boundary region of the set X as follows:

$$\underline{B}X = \{x|\mu_X^B(x) = 1\} \tag{D.6}$$

$$\overline{B}X = \{x|0 < \mu_X^B(x) \leq 1\} \tag{D.7}$$

$$BN_B(x) = \{x|0 < \mu_X^B(x) < 1\} \tag{D.8}$$

$$U - \overline{B}X = \{x|\mu_X^B(x) = 0\}$$

These definitions are consistent with those given in Section D.3.

In addition, using this membership function, we can also define an arbitrary level of precision $a \in [0,1]$ of the lower- and upper approximations:

$$\underline{B}_aX = \{x|\mu_X^B(x) \geq a\} \tag{D.9}$$

$$\overline{B}_aX = \{x|\mu_X^B(x) > 1 - a\} \tag{D.10}$$

D.5   DEPENDENCY OF ATTRIBUTES

In data analysis, it is important to discover dependencies between the condition and decision attributes. Using these dependencies, one can identify and omit the unnecessary attributes (which are considered to be redundant information in the decision system) and the corresponding values for making decision or classification. Intuitively, we say that a set of decision attributes Td depends totally on a set of condition attributes Tc, denoted as Tc $\Rightarrow$ Td, if all the values of decision attributes are determined uniquely by values of the condition attributes. This implies that there exists a functional dependency between values of Tc and Td. Note that the dependency can also be partial, which means that only some values of Td can be determined by values of Tc. Now we give its formal definition:

Definition D.7. Let C and D be subsets of A, $D \cap C \neq \emptyset$ and $D \cup C = A$. We say that D depends on C in a degree k $(0 \leq k \leq 1)$, denoted $C \Rightarrow_k D$, if

$$k = \gamma(C,D) = \sum_{X \in U/D} \frac{|\underline{C}X|}{|U|} \tag{D.11}$$

where $|X|$ indicates the cardinality of X and U/D denotes the partition determined by D (i.e., the family of all equivalence classes of I_D).

If $k = 1$, we say that D depends totally on C, and if $k < 1$, we say that D depends partially (with a degree k) on C. The degree k, called the *degree of the dependency*, means the ratio of all elements of the universe that can be properly classified into the partition U/D employing attributes from C. For example, in the decision system illustrated in Table D.2, the attribute Pf depends on the set of attributes $\{E,Q,L\}$ of degree $\frac{2}{3}$. This is explained below.

$$C = \{E,Q,L\}, D = \{\text{Pf}\}, U/D = \{\{1,3,6\}, \{2,4,5\}\},$$

if $X = \{1,3,6\}, \underline{C}X = \{1,6\}$;

if $X = \{2,4,5\}, \underline{C}X = \{4,5\}$.

$$k = \gamma(C,D) = \sum_{X \in U/D} \frac{|\underline{C}X|}{|U|} = \frac{|\{1,6\}|}{|\{1, 2, \ldots, 6\}|} + \frac{|\{4,5\}|}{\{1, 2, \ldots, 6\}|} = \frac{2}{3}$$

This means that only four of the six objects in U (i.e., $\{1,4,5,6\}$) can be identified exactly by using the attributes E, Q, and L. Here the decision attribute Pf depends partially on the condition attributes E, Q, and L, of degree $\frac{2}{3}$.

REFERENCES

1. A. Marek and Z. Pawlak, Rough sets and information systems, *Fundamenta Informaticae*, vol. 17, pp. 105–115, 1984.

2. Z. Pawlak, Rough sets, *International Journal of Computer and Information Science*, vol. 11, pp. 341–356, 1982.

3. Z. Pawlak, *Rough Sets: Theoretical Aspects of Reasoning about Data*, Kluwer Academic, Dordrecht, The Netherlands, 1991.

4. Z. Pawlak, Rough set theory for intelligent industrial applications, in *Proceedings of the Second International Conference on Intelligent Processing and Manufacturing of Materials (IPMM-99)*, Honolulu, HI, IEEE, Press, Piscataway, NJ, vol. 1, pp. 37–44, 1999.

5. D. Dubois and H. Prade, Rough fuzzy sets and fuzzy rough sets, *International Journal of General Systems*, vol. 17, pp. 191–209, 1990.

6. D. Dubois and H. Prade, Similarity versus preference in fuzzy set-based logics, in *Incomplete Information: Rough Set Analysis* (E. Orlowska, ed.), Physica-Verlag, Heidelberg, pp. 440–460, 1998.

7. A. G. Jackson, M. Ohmer, and H. Al-Kamhawi, Rough sets analysis of chalcopyrite semiconductor band gap data, in *Proceedings of the Third International Workshop on Rough Sets and Soft Computing (RSSC-94)*, San Jose, CA, pp. 408–417, 1994.

8. A. G. Jackson, S. R. Leclair, M. C. Ohmer, W. Ziarko, and H. Al-Kamhwi, Rough sets applied to material data, *Acta Materialia*, vol. 44, no. 11, pp. 4475–4484, 1996.

9. P. Lingras, Rough neural networks, in *Proceedings of the Sixth International Conference on Information Processing and Management of Uncertainty in Knowledge-Based Systems*, Granada, Spain, Springer-Verlag, Berlin, vol. 2, pp. 1445–1450, 1996.

10. A. Mrozek, Rough sets in computer implementation of rule-based control of industrial processes, in *Intelligent Decision Support: Handbook of Applications and Advances of the Rough Set Theory* (R. Slowinski, ed.), Kluwer Academic, Dordrecht, The Netherlands, pp. 19–31, 1992.

11. R. Nowicki, R. Slowinski, and J. Stefanowski, Rough sets analysis of diagnostic capacity of vibroacoustic symptoms, *Journal of Computers and Mathematics with Applications*, vol. 24, no. 2, pp. 109–123, 1992.

12. J. F. Peters, K. Ziaei, and S. Ramanna, Approximate time rough control: concepts and applications to satellite attitude control, in *Proceedings of the First International Conference on Rough Sets and Current Trends in Computing*, Warsaw, Poland, Springer-Verlag, Berlin, pp. 491–498, 1998.

13. L. Polkowski and A. Skowron (eds.), *Rough Sets in Knowledge Discovery*, Physica-Verlag, Heidelberg, vol. 1, no. 2, 1998.

14. A. Skowron and C. Rauszer, The discernibility matrices and functions in information systems, in *Intelligent Decision Support: Applications and Advances of the Rough Sets Theory* (A. Slowinski, ed.), Kluwer Academic, Dordrecht, The Netherlands, pp. 331–362, 1992.

15. R. Slowinski, *Intelligent Decision Support: Handbook of Applications and Advances of the Rough Set Theory*, Kluwer Academic, Dordrecht, The Netherlands, 1992.

16. R. Slowinski, Rough set approach to decision analysis, *AI Expert*, vol. 10, pp. 18–25, 1995.

17. R. Slowinski, Rough set theory and its applications to decision aid, *Belgian Journal of Operations Research, Special Issue Francoro*, vol. 35, no. 3–4, pp. 81–90, 1995.

18. J. Wróblewski, Finding minimal reducts using genetic algorithms, in *Proceedings of the International Workshop on Rough Sets and Soft Computing in conjunction with the Second Annual Joint Conference on Information Sciences*, Wrightsville Beach, NC, pp. 186–189, 1995.

19. J. Wróblewski, Genetic algorithms in decomposition and classification problems, in *Rough Sets in Knowledge Discovery 2: Applications, Case Studies and Software Systems* (L. Plokowski and A. Skowron, eds.), Physica-Verlag, Heidelberg, pp. 472–492, 1998.

20. S. K. Pal and A. Skowron (eds.), *Rough-Fuzzy Hybridization: A New Trend in Decision Making*, Springer-Verlag, Singapore, 1999.

INDEX

Foundations of Soft Case-Based Reasoning. By Sankar K. Pal and Simon C. K. Shiu
ISBN 0-471-08635-5 Copyright © 2004 John Wiley & Sons, Inc.